CHAUCER'S

TREATISE ON THE ASTROLABE.

CHAUCER'S

TREATISE ON THE ASTROLABE.

THE TREATISE

ON THE ASTROLABE,

OF

EDITED

WITH NOTES AND ILLUSTRATIONS,

BY

ANDREW EDMUND BRAE.

"And to his sonne that called was Lowys
He made a treatise full noble and of gret prise."
LYDGATE.

LONDON:
JOHN RUSSELL SMITH,
36, SOHO SQUARE.
MDCCCLXX.
1 8 70
107

CONTENTS.

PAGE.

Introduction 1
Explanation of the Plates 16

The Proheme 19
Description of the Astrolabe 22

The Conclusions.

I. The Sun's Place 32
II. Taking Altitudes 33
III. The Day of the Month 33
IV. The Hour of Day or Night and the Ascendant . . 33
V. The Technical Ascendant 35
VI. The Mean of Almicanters 37
VII. Twilight 37
VIII. The Diurnal Arke 38
IX. The Conversion of Hours 38
X. The Vulgar Day 38
XI. Hours Inequale by Day and Night 39
XII. Equal Hours 39
XIII. Planetary Hours 40
XIV. Oblique Ascension 41
XV. Declination of a Point in the Ecliptic . . . 41
XVI. Latitude of Construction 41
XVII. Meridian Altitude 42
XVIII. Sun's Place in the Ecliptic 42
XIX. Similar Days 42
XX. Similar Points in the Ecliptic 43
XXI. Stars Indeterminate 43
XXII. Stars Determinate 44
XXIII. Elevation of the Pole 45
XXIV. Terrestrial Latitude 45
XXV. Altitude of the Pole 46
XXVI. Terrestrial Latitude 47
XXVII. Right and Oblique Ascensions 48
XXVIII. Right Ascension of Signs 48
XXIX. Oblique Ascension of Signs 49
XXX. The Cardinal Points 50
XXXI. Celestial Latitudes 50

		PAGE.
XXXII.	Azimuth at Rising	51
XXXIII.	Points of the Compass	52
XXXIV.	Bearing of a Conjunction	52
XXXV.	Azimuth in Altitude	53
XXXVI.	Celestial Longitudes	53
XXXVII.	Motion Direct or Retrograde	53
XXXVIII.	Equations of Houses	54
XXXIX.	Equations of Houses	55
XL.	Meridional Line	55
XLI.	Terrestrial Longitude	56
XLII.	(Apochryphal ?)	57
	Practice of Umbra Recta and Umbra Versa	59

APPENDIX.

REPRINTS.

I.	The Pilgrimage to Canterbury	65
II.	The Arke of Artificial Day	68
III.	Astronomical Evidence	71
IV.	The Star Min Al auwâ	74
V.	Tests of Positions	79

NOTES ON THE REPRINTS.

A.	The Halfe Cours in Aries	81
B.	The Zodiacal Signs	84
C.	The Angle Meridional	86
D.	The Star Min Al auwâ	88

ESSAYS.

ON THE MEANING OF CHAUCER'S PRIME	90
ON THE CARRENARE	101
ON SHIPPES OPPOSTERES	106

ERRATA.

Page 5, line 2, for "1798" read "1598."

Page 56, Conclusion XL.—The alteration made in the text of this Conclusion, as explained in the foot-note, is unnecessary. For the apparatus may have been intended for the Winter and not for the Summer Solstice, in which case the original proportion of one quarter of the shadow for the gnomon would be sufficiently correct.

Page 45, in foot note, for "pp. 23, 24," read pp. 9, 10.

Page 88, in Note D, *passim*, for "min al away," read min al auwâ.

Page 91, line 30, for "cœval," read coeval.

Page 94, line 20, for "jeutaculum," read jentaculum.

PRINTER'S AND OTHER ERRORS.

Page Line

5 2 For 1798 read 1598.

21 6 For one to one read *owre to owre.*

31 2 Insert the following paragraph, which is in every edition, printed and MS., but was accidentally omitted here—

"And everich of these signes hath respecte to a certeyn parcell of the bodie of a man and hath it in governaunce. As Aries hath thyn hed, and Taurus thy necke and thy throte, Gemini thyne armholes and thyne armes, and so forth as shall be showed more plain in the fifte partye of this tretys. The Zodiak the which is partye of the eyght spere overkervith the equinoctial and he everkervith him ageyn in evene partes, and that one half declineth southward and that other northward as plainlie declareth the tretys of the spere."

33 21 For "when *thou* liste," read 'when *thee* liste'

[But see 'Knight's Tale' 1183, where the Corpus MS. has '*thou list*' and the Lansdowne MS. the same.]

40 4 Transpose, and read "the laste chapter of the fowerth partye."

40 In foot-note add, "But it must be said that Steven's 'teching' is supported by several MSS."

56 21 For 'signet' read 'line meridional.'

62 9 For *nere* read *further.*

And for the first note substitute

"Stevens properly changed *nere* of the MSS. into *further.*"

In the 6th line of the second footnote deme "*the two first of the terms are transposed,*" and substitute "*the difference of the points multiplied by the length of the scale is the first term, the second is the distance between the stations, the third is the product of one point multiplied by the other, and the fourth is the altitude required.*"

90 23 For 'John de Belethus' read 'Joannes Belethus.'

94 20 For 'jeutaculum' read jentaculum.

99 22 For '*Thomas*' read *Francis.*

Introduction.

ALTHOUGH the Treatise on the Astrolabe is, in some respects, the most interesting of Chaucer's works—inasmuch as it brings us into familiar and almost domestic communion with his individual self, while he describes to his "lytel sonne," with delightful simplicity and in the most inartificial language, the sort of scientific knowledge which in those early days, even more than at present, was considered necessary to a gentleman's education—yet it has received so little care and attention from the editors of his works, that, since the edition of Urry, in 1721, it has not been included in any modern reprint. And even then, Urry did little more than blindly copy from preceding editions, without any attempt to explain, amend, or illustrate the text.

Several years ago (in 1851) I published a series of papers explanatory of the astronomical allusions of Chaucer in the Canterbury Pilgrimage. These I shall reprint in an Appendix to this volume—if for no better reason— to shew the length of time my attention has been given to the subject, as well as to rescue the papers themselves from the oblivion of ephemeral publication. In the preparation of those papers I had necessarily recourse to Chaucer's Treatise on the Astrolabe, as printed in Urry's edition of the works, and I found it in such a deplorably faulty and neglected state, that the necessity of rectifying such parts as I then required awakened in me a desire to re-edit the whole. And although various causes had prevented the fulfilment of this object until the present time, it had never been lost sight of. For it appeared very evident to me then—and I have seen no reason since to alter the opinion—that a direct connection may be traced between the subject I was then engaged upon, namely, the astronomical evidence of date in the Canterbury Tales, and this Treatise on the Astrolabe. The date of the Pilgrimage, as I then endeavoured to shew, was 1388, and the avowed date of this Treatise is 1391 : it seems then an almost unavoidable inference,—that we owe this practical treatise to the preparative study of the subject undertaken by Chaucer for the purpose of inventing and verifying his intended astronomical phenomena in diversifying the incidents of his Canterbury Pilgrimage.

And yet this interestieg attempt on the part of an imaginative poet to render science familiar in what may be termed one of our earliest and simplest popular treatises, is that portion of Chaucer's works which, of all others, his editors have most culpably misunderstood and neglected.

Nor is it in the editing only of former printed editions of this little treatise that it appears to have been read with strange absence of care and intelligence; the same fate has attended it when only incidentally alluded to. For example, J. Thompson, in his "Life of Geoffry Chaucer," prefixed to Urry's edition of the works, writes :—

"He (Chaucer) retired to Woodstock; and weary of a long series of hurry, noise, danger, and confusion, he shifted it for quiet and the calm pleasure of a studious safety, which produced his excellent Treatise of the Astrolabe, which is calculated for the latitude of Woodstock, being a small matter different (as he says) from that of Oxford."

This last assertion "*(as he says)*" Thompson attempts to support by referring in a foot note to this passage in Chaucer's text, *infra*, page 47.

"I suppose that the sonne is thilk day at none 38 degrees of heyght; abate then 38 degrees out of 90, so leveth ther 52; then is 52 degrees the latitude. I saye not this but for ensample, for wel I wote the latitude of Oxenforde is certayne minutes lesse."

It seems hardly credible that any person of average intelligence could, even in carelessness, so misapprehend this expression, "certaine minutes lesse," when it is so plain that it is Chaucer's reservation for having expressed the latitude of Oxford in round numbers as 52 degrees, instead of that more exact latitude of fifty-one degrees and fifty minutes he elsewhere assigns to it. And yet, in the face of this, and in the face of Chaucer's express declaration to his son that he presents him with "A sufficient astrolabie compowned after the latitude of *Oxenforde*," his biographer gravely declared, as a historical fact, that this same Astrolabe was calculated for the latitude of Woodstock.

I have met with another reference to Chaucer's astrolabe, which might have been valuable as an illustration had it not been vitiated by similar mis-reading. It occurs in a paper by Mr. Burrow, in the Appendix to the second volume of Asiatic Researches (anno 1789), "I compared an astrolabe in the Nagry character (brought by Dr. Mackinnon from Jynagur) with Chaucer's description, and found them to agree most minutely; even the centre-pin, which Chaucer calls 'the horses,' had a horse's head upon it." Now Chaucer does not call the centre-pin "the horses"; what he says is—"thorowe which pin ther goeth a lytel wedge which is cleped the Horse." So that this comparison with the Indian Astrolabe, which might have been so interesting, becomes of no value, in the uncertainty as to what part of it it was in which Mr. Burrow really did see the horse's head.

Whether the wedge-like form of a horse's head was given to this cross-pin, and thereon the likeness further engraved by way of ornament, or whether some other fancy suggested the name, it is certain it was very generally given to it; the Arabs called it *Al-pheras*, a name also given to the knight's-piece at Chess, which, as everyone knows, is but a horse's head. The use of the wedge, when thrust through the great centre-pin, was to twitch up and render immovable during observation certain adjustable plates in the interior of the instrument. Through these plates the centre-pin passed, and was jammed against them by the wedge: but it is not clear from Chaucer's description how this was done, nor on which side of the plate the wedge was applied. And, in addition to these difficulties, it cannot be supposed that a loose pin, merely wedged in, would afford anything like an invariable centering for a radial index, on which so much of the correctness of such an instrument would depend, On these points I shall hazard a conjecture when I come to describe the several details.

The adjustable plates just mentioned are called by Chaucer "the plates for divers clymates"—*i.e.*, for divers latitudes: of which that particular plate required for the observer's locality would be selected; and being placed uppermost and adjusted to a proper position, would be there secured by the insertion and jamming of the centrepin. Thus it would become "*the plate under the rete*," as Chaucer calls it when describing its construction for the latitude of Oxford.

It is interesting to find these plates confirmed and illustrated in "The Arabian Nights," by mention of the Barber's astrolabe "*of seven plates.*" But this is only in Mr. Lane's translation, the corresponding description in Galland's version being "un astrolabe bien propre;" and in "The Book of 1000 Nights and One Night"—a translation by Henry Torrens published in Calcutta in 1838, the same passage is rendered—"*An astrolabe and it had seven sides mounted with silver.*" Thus confirming in some degree Mr. Lane's translation; which, however, is greatly to be preferred, coinciding as it does so closely with Chaucer's description. Mr. Lane, moreover, renders his translation still more valuable as an illustration, by the following note to the passage:—

"The astrolabe is more commonly used by the Arabs than any other instrument for astronomical observations. It is generally between four and six inches in diameter. It consists of a circular plate with a graduated rim, within which fit several thinner plates; and of a limb, moving on a pivot in the centre, with two sights. The plates are engraved with complicated diagrams, &c., for various calculations. The instrument is held by a ring, or by a loop of cord attached to the ring, during an observation: and thus its own weight answers the same purpose as the plumb line of the quadrant (which the Arabs sometimes use in its stead): the position of the moveable limb, with the sights, marking the required altitude."—*Lane's Arabian Nights*, chap. v., note 57.

Here again it is to be regretted that the writer of this description of the Arabian astrolabe had not a more practical knowledge of such instruments than is evinced by his calling the index *a limb:* because the term limb, being derived from *limbus*, is practically appropriated to the circular rim, and never to the radial index. He would then, doubtless, have given a more detailed and more reliable account; and would not have fallen into the strange mistake, in another note to the same passage, of asserting :—

"A degree is four minutes; it would have been more proper, therefore, to have said eight degrees and two minutes, than seven degrees and six minutes." —(*Ibid.*, note 60).

This note seems to have proceeded from some unaccountable mental confusion between degrees of arc and minutes of time: and, although Mr. Lane makes no allusion to Chaucer or his astrolabe, yet this division of a degree into four minutes might seem curiously enough to have had Chaucer for its authority in the following passage of this treatise—"and every degre of this bordure conteineth foure minutes." But then Chaucer takes good care to add—"that is to saie foure minutes of an hour"—a necessary distinction. I do not mention these slight mistakes in a spirit of criticism, but simply because it would not have been possible to quote these notes at all without noticing in some way these mistakes in them. When Chaucer wrote "every degree containeth foure minutes," he was comparing together two different accounts—but when Mr. Lane wrote, "A degree is four minutes," it is plain from the context that he meant in one and the same account.

In describing the degrees of the outer border Chaucer again inculcates the necessary distinction between time and arc :—

"And I have said five of these degrees maken a mile-waie and thre mile-waie maken an hour: and every degre of this border conteyneth foure minutes, and every minute fourtie secondes."

Here *fourtie* is a palpable error, the correction of which to *foure* renders the meaning perfectly plain—*i.e.*, every fifteen degrees of the outer border contain (or are equivalent to) one hour of the inner (or time circle): every degree of the outer border is equal to four minutes of the inner: and every minute of the outer to *four seconds* of the inner.

There can be no doubt that this is the substance of what Chaucer wrote; yet in all the MSS. that I have examined, and in all the printed copies previous to Speght's second edition in 1602, the last five words are—"and every minute lx secondes." In Speght's second edition, which professed to be an amended copy of the first, this "lx" was changed into fourtie; a still greater error, which was copied into 1687, and into Urry's edition of 1721. If the change had been from lx to xl, both being in roman numerals, it would appear to have been a mere accidental transposition of the numerals, but that explanation is ırred by the substituted number being printed at length "fourtie." Therefore the only way to account for the change is to suppose that Speght, in the interval between 1798 and 1602, may have seen some MS. with the correct reading "foure," and converted it into fourtie.

Another and a far more unaccountable error of the printed copies occurs in Chaucer's VIIth Conclusion :—

"The nadyre of the sonne is thylk degre that is opposyte to the degre of the sonne in the 320 signe."

So in Thynne's, and in all other printed copies previous to Speght's first edition of 1598 ; in which, for some incomprehensible reason, he printed "in the xxiii signe" ! But in his second edition of 1602—as if he regretted his departure from the older copies—he restored, in his amended text, "the 320 signe," in which he was again followed by 1687, and by Urry in 1721.

Now, this changing and rechanging show an amount of *deliberation over this error* that renders its repetition and persistence doubly astonishing. Because even if these editors had been ignorant of such a common place fact as that there are only twelve signs altogether in the zodiac, they might have learnt it from Chaucer in the very work they were engaged in editing. They all boast of having compared their text with many MSS., and yet there is no manuscript —none, at least, that has come under my observation—that would not in this case have given the right reading, which, it is scarcely necessary to say, is "in the VII signe."*

Although there are many manuscripts in the British Museum containing poetical works of Chaucer. I could discover only three in which this prose Treatise on the Astrolabe is to be found. Of these, that which is apparently the most ancient is numbered 23002 (additional MSS.). It is a comparatively recent acquisition, having been purchased for the Museum at Mr. Dawson Turner's sale in 1859. It does not differ greatly from the other MSS., except in the omission of some portions and the transposing of others. Altogether although it may assist the text in some few cases, it cannot be looked upon as a faithful or reliable copy.

The second is a Sloane MS. 314, chiefly remarkable for an assertion on the fly-leaf that it is in the hand-writing of Chaucer himself! And this was so far credited by the Rev. Samuel Ascough, who compiled the Catalogue of the Sloane Manuscripts in the British Museum, that he copied the assertion in his catalogue and added—"*if it is not in Chaucer's own writing, it is nearly of the same age.*"—But in this opinion he was surely mistaken : for, if there were no other refutation of it, the rude and ill executed diagrams which are incorporated with the text of this MS. are evidently copied from Stoeffler's book, *De Fabrica Astrolabii*, which did not appear till the year 1513 (see *infra*, in description of

* "Nam in septimo signo fit solstitium a bruma : in septimo bruma a solstitio : in septimo equinoctium ab equinoctio."—AULUS GELLIUS, iii., 10.

Plate II, the note upon Chaucer's Scale of Umbra Recta and Versa, which is single, while that of Stoeffler is double; and I take that to be one test by which copies from Stoeffler may be detected.)

The third is also a Sloane MS., No. 261, and is very interesting, in so much that although never printed it was evidently written and prepared for the press with a view to publication. It has a regular title, dedication, and address to the reader: emendations of, and comments upon the text; and it is illustrated by pen-and-ink diagrams similar to those in 314, and certainly also derived from Stoeffler's book. But the most remarkable circumstance attending these two manuscripts is that 314 was obviously in the possession of him who wrote 261—probably the very original from which he copied it. For on the margins of 314 are corrections and remarks in the handwriting of 261: or, if the identity of the hand writing be disputed, then there are references in the one to certain alterations made in the text of the other which place the fact beyond all reasonable doubt. From these circumstances attending its preparation the MS. 261 possesses almost the authority of a printed book zealously edited; and indeed it is very much more correct than any of the printed copies. And since a comparison with it would unquestionably have corrected some of the more glaring errors of Urry's edition, it is no slight reproach to him that he not only had this MS. in his hands, but he left a record of that fact by inscribing on the fly-leaf this autograph memorandum "This MS. belongs to Dr. Sloane, T. Urry, 1709." The title of this MS. is "*The Conclusions of the Astrolabie compyled by Geoffry Chaucer newlye amendyd.*" But while it has every other requisite for a printed book, it is singular that no date is any where affixed to it; and as that is a very interesting particular I shall be at some pains to establish it from the Dedication, which is as follows:—

"To the righte honorable and his verrie good lorde Edwarde earle of Devōshier Walter Stevins wissheth continuall encrease of honoure."

"When I had amended this lyttell worke (my right honorable lorde) which I now offer unto yower good lordship; me thought my labours therein wolde not onely be allowed of as manye as delyte in those wyttie conclusions of the Astrolabie, but also not unthankefull to your exceeding wisdome; who amongst your other vertuous studies hath had a great felicitie and pleasure in these mathematicall practises also; which thinges especiallie moveth me to dedicate this my lytell travaile unto youre lordship. And although I confess this to be so slender and simple an offer for youre excellencye, yet it may please you to take it as an argumente of the faithfull good will and humble harte I owe to your honour which God advance and encrease even as your vertues and worthiness right well deserve."

Now it so happens that this dedication to Edward Earl of Devonshire affords a very close determination of the date. There are only three noblemen, all of the House of Courtency, to whom the name Edward, and the title Earl of Devonshire could apply.

The first Edward succeeded to the title upon the death of his grandfather Hugh in 1419.

The second Edward received the title from Henry VII after the battle of Bosworth. He survived till 1509; but could not have been the object of the Dedication, since Stoeffler's book, which is referred to in Stevins' MS., was not written till 1510, nor printed till 1513.

The third Edward was that unfortunate young nobleman, the great grandson of the preceding, whose father Henry, after having been created Marquis of Exeter by Henry VIII, was attainted and executed in the same reign. Thereupon this son Edward, then a boy of twelve years old, was imprisoned and so remained during his whole youth and early manhood, until released and restored in blood by Queen Mary in 1553. But even then his liberty was of short duration, for in little more than two years he was again imprisoned and only finally released to go into a sort of voluntary exile wherein he died at Padua in 1556. He does not appear to have ever borne his father's title of Marquis of Exeter, although he is so styled by Collins, Courthope, and others; and it is necessary to the complete identification of the dedication to show that these authorities are wrong. In contemporary chronicles he is invariably called Earl of Devonshire: and in the funeral oration pronounced at his obsequies in Padua by Dr. Thomas Wylson, wherein, if ever, all the honours and dignities to which he had been entitled would be declared—and wherein his parents were severally mentioned as Marquis and Marchioness of Exeter; *he* is spoken of all through as Comes Devoniæ. Thus, when the favours are enumerated which Queen Mary had conferred upon him, they are thus particularized: —

"Libertatem donavit, decus restauravit, et ad dignitatem summam evexit, sic ut pientissima Reginæ opera Comes Devoniæ ab omnibus salutaretur."

But the most conclusive proof that this Edward was never Marquis of Exeter is found in the Calendar of State Papers, Domestic Series, edited by Mr. Robert Lemon. In this there is a list of upwards of 120 letters to and from this Edward Courtnay during his absence abroad, up to his death at Padua; and in all these letters he is styled by himself and by others Earl of Devonshire. Some of these letters are from his own mother, the Marchioness of Exeter, and upon one, dated Nov. 8, 1555, there is this autographic address:

"To my son the erle of Deffonsher thys be delyvred."

Now, besides showing that the title Earl of Devonshire was the proper one for a dedication to this young nobleman, it has been desirable to establish it for another reason, and that is, that it narrows the date of the dedication to the short interval of three years between his creation to the title of Earl of Devonshire in 1553 and his death in 1556. Because, had the dedication been written *before* his legal restoration, while yet a prisoner, the title given to him then by his dependents would be one of compliment and courtesy, and in that case it would assuredly have been the title his father had borne—Marquis of Exeter.

It only remains to establish the probability, or indeed the certainty, that it was to this young nobleman Walter Stevins dedicated his Manuscript of the Conclusions of the Astrolabie. Of this there are three very sufficient presumptive proofs:

1. The mention of Stœffler's book, which limits the date in one direction to 1513; and the absence of any Earl of Devonshire named Edward after 1556 which limits it in the other.
2. The probability derived from the studious character of this young nobleman, of whom Dr. Wylson, in the funeral oration already mentioned, spoke in these terms:—

   "Nec angustia loci, nec solitudo, nec amissio libertatis illum a litteris avocaret. Unde tam avide philosophiam arriepat, et tantas in ea progressiones faciebat, nemo ut illi ex principibus par esset. Neque in hoc solum laudabili studio seipse exercuit, sed intima naturæ scrutatus mysteria: mathematicorum labyrinthum intravit fructu summo et voluptate singulari."

   A curious resemblance may be observed in these concluding words to a passage in the dedication—"who, amongst your other vertuous studies, hath a great felicitie and pleasure in those mathematical practises also."
3. The following note subjoined in the MS. to that passage in Chaucer's treatise wherein the vernal equinox is attributed to the 12th March. "Albeit in Chaucer's time upon the 12 daie of March the sonne entred into the hedde of Aries, yet in oure time you shall finde that the sonne entreth therein on the 10th daie of the same moneth."

It so happens that there is a rather curious evidence that the 10th March was generally received about the middle of the sixteenth century as the day of the vernal equinox. Puttenham, in his "Arte of English Poesie," treats the 10th March as so *notoriously* identified with the vernal equinox, that he censures a poet's particular assignment of that day to it as an unnecessary redundancy. "For if," says he, "the thing or person they go about to describe by circumstance be by the writer's improvidence otherwise bewrayed, it loseth the grace of a figure, as he that said:

"The tenth of March when Aries received
Dan Phœbus' raies into his horned head"—

' intending to describe the spring of the yeare which every man knoweth of himself hearing the day of March named. The verses be very good, the figure nought worth, if it were meant in periphrasis for the matter. That is, the season of the yeare which should have been covertly disclosed by Ambage was by and by blabbed out by naming the day of the moneth."

Lib. iii. "Of Ornament."

Puttenham, as was too often his wont, gives no reference or clue by which the quotation thus criticised could be identified; but with a little research I have ascertained that it is one of George Gascoigne's poems, and may be found in his "Hearbes," printed about 1575.* Now, unless the association of the equinox with the 10th of March had been very notorious, and of long standing there would have been no point in Puttenham's criticism. The question here is not what really was the actual day of the equinox in any given year, but what was popularly and generally supposed to be the day.

These accumulated proofs preclude all reasonable doubt that Walter Stevins wrote this MS. about 1555, and dedicated it to Edward Earl of Devonshire, the last of those names. The views and pretensions of the writer will best appear in his own words in an address which immediately follows the dedication already quoted, and which is also worth transcribing, for the additional reason that many of the remarks made in it upon the state of the text of the Treatise on the Astrolabe, in the writer's time, are as applicable to it now as they were three centuries ago :—

"When I happenyd to look upon the conclusions of the Astrolabie compiled by Geffray Chaucer and founde the same corrupte and false in so many and sondrie places that I doubted whether the rudeness of the worke were not a greater sclander to the authour than trouble and offense to the readers, I dyd not a lytell mervell if a book should come oute of his handes so imperfect and indigest whose other workes weare not onely rekoned for the best that ever weare set forthe in oure english tonge, but also weare taken for a manifest argument of his singular witte and generalitie in all kindes of knowledge. However be it when I called to remembrance that in his prohem he promised to sette forthe this worke in five partes, whereof weare never extante but those two first partes onely, it made me to believe that either the work was never fynisshed of the authour, or els to have ben corrupted sens by some other meanes, or what other thynge might be the cause thereof, I wiste not. Never the lesse understanding that the worke, which before lay al neglected, to the profit of no man and discourage of many, might be tourned to the commoditie of as manye as hereafter should happen to travayl in that parte of knowledge I thowght it a thinge worthe my laboure if I could sette it in better order and frame. Which thinge however I have done it let be thine indifferent judgement which heretofore have readen th' other setting forth ; or liste to repaire this and that together ; wherein I confess that besides the amendinge of verie many words, I have displaced some conclusions, and in some places where the

---

* "This tenth of March, when Aries receyu'd
Dan Phœbus rayes into his horned head :
And I my selfe by learned lore perceyu'd
That Ver approcht and frostie winter fled ;
I crost the Thames, to take the cherefull ayre
In open feeldes, the weather was so fayre," &c.

sentences weare imperfite I have supplied and filled them as necessitie required. As for some conclusions I have altered them, and some have I cleane put out for utterly false and untrew; as namely the conclusion of direction and retrogradacion of planetes and the conclusions to know the longitudes of starres whether thei be determinate or indeterminate in the Astrolabie. The conclusion to know with what degree of the zodiack any planet ascendeth on the horizon whether his latitude be northe or southe—as the meanyng of the same conclusion was most hardeset by reason of the imperfitenes thereof, so in practise I found him most false—as he shall find that list to take the like paines. Notwithstanding, this have I done not challenging for myself but revising and leaving to worthie Chaucer his one praise for this worke, which if it had come parfite into oure handes (no doubt) would have merited wonderfull praise. As for me if I have done any thing therein it shall suffice if the lovers of wittie Chaucer do adopt my good will and entente. VALE."

It will be seen in the course of the present reprint that I do not by any means adopt all the alterations and additions made by Stevins. But as his zeal in the work is unquestionable, and as he has judiciously corrected many of the more glaring errors of the printed editions (of which no less than three—1532, 1542, and 1545—had been published shortly before he prepared his MS.), his corrections are worthy of being noted in the margins.

Of the conclusions specially mentioned by him towards the close of the foregoing address, namely: "The conclusion of direction and retrogradacion of planetes" (XXXVIIth of this edition), and "the conclusions to know the longitudes of sterres whether thei be determinate" (XXII) "or indeterminate" (XXI)—he has omitted the first and last,—or, as he expresses it, "cleane put out for utterly false and untrew"—but he has inadvertently retained XXII, although coupled by himself with XXI in the same sentence. He also specially mentions "The conclusion to know with what degree of the zodiack any planet ascendeth on the horizon whether his latitude be northe or southe" (XLII); and this he retains and endeavours to render practicable by certain additions and alterations of his own devising.

Now it seems probable that Stevins, in dealing with these Conclusions, was not actuated by any judgment of his own, but was influenced by Stœffler's book, with which he was well acquainted. It is certain that the three Conclusions declared by him to be "utterly false and untrew" are three out of seven propositions which Stœffler especially condemns. These three, as numbered in his list of seven, are:—

"IV. Inquirere an planeta sit directus aut anomalus sive retrogradus."

"V. Perscrutari in quo gradu signi sit quælibet stella fixa, *in reti descripta* (i.e., *determinate*).

"VII. Determinare signum et gradum cujuslibet stellæ fixæ, *in aranea non positæ*" (i.e., *indeterminate*).

Now these are, almost literally, Chaucer's XXXVIIth, XXIInd, and XXIst, while his XLIInd, which Stevins retains, is not expressly denounced by Stœffler.

My opinions respecting these Conclusions are stated in the notes subjoined to them. That which Stevins has retained (the XLIInd) is the only one that I absolutely repudiate and deny to be Chaucer's. It is quite foreign to his manner of going to work in his other conclusions; and he expressly defers, in his Proem, the subject of this problem to the fourth part of this treatise: which fourth part was either too laborious a task for him to complete, or, if completed, was afterwards lost. The object of the XLIInd conclusion, the last of the astronomical series, is:—

"To knowe with what degree of the Zodiake that any planet ascendeth on the orizonte whether his latitude be north or south."

And at the end of the conclusion there is a special instruction to be observed in finding "the arysing of the Mone." Now Chaucer, in his Proem, says:—

"The fourthe partye shall be a theorike to declare the mevyng of the celestiale bodyes, with the causes: the which fourthe partye in special shal shewe, in a table of the very mevynge of the Moone, from one to one, every daye and every signe, after thin almanacke. Upon the which table there followeth a canon sufficyent to teche, as wel in maner of workynge in the same conclusions, as *to knowe in oure orizonte with which degre of the Zodiac the Mone ariseth in any latytude, and the arysynge of any planete after his latytude fro the eclyptike lyne.*"

Now, it may be observed that these concluding words are almost identical with the XLIInd Conclusion; and it is not improbable that that conclusion was invented and interpolated in after times in an attempt to realize this problem by the Astrolabe, independently of the tables to which it was relegated by Chaucer. Had Chaucer himself attempted to do this, he would never have had recourse to the clumsy machinery described,—which, I repeat, is quite foreign to *his* manner of going to work. There is no scale of latitude—"fro the eclyptike lyne"—upon the Astrolabe; and it is not stated in the problem whence the distance of the compass points is to be obtained: if from the meridional almicanters, there would be no necessity for the points of the compasses and the waxed label, for Chaucer would have arrived at the same result in a far simpler way; he would have brought the degree of the planets' longitude to the meridional line (as in Conclusion XV), and finding the given latitude, above or below the ecliptic, in the almicanters, he would lay the label on it and mark it thereon with a "prycke of inke:" he would then "*turn the rete about joyntly with the label,*" and bringing the mark to the horizon would observe with what degree of the zodiac it would arise. But Chaucer seems to have been quite aware that such a process would not give the true place of the planet (although it would be just the same as that obtained by the ponderous method described in XLII); and, therefore, he properly reserved this conclusion for determination by tables of planetary motions.

When the period at which Chaucer wrote is taken into consideration, and that after all he was but an *amateur* astronomer, his general correctness is something admirable. The latitude assigned by him to Oxford differs by scarcely more than four minutes from its rigid determination at the present time as noted in the Nautical Almanac: and so sensitive was he on the score of its correctness that when, for the sake of brevity, he had increased it by only ten minutes, he takes care to explain his having done so—"for wel I wote the latitude of Oxenforde is certayne minutes lesse." (See *ante*, page 2) Compare this with a professed and pretentious astronomer (Stoeffler) who, a hundred years later, in his "*Tabula Regionum, Provinciarum, et Oppidorum Insigniorum Europæ*" (wherein, by the way, he ignores London altogether), makes both latitude and longitude of Oxford more than a degree too much: or with Stoeffler's pupil, Sebastian Munster, the cosmographer, who declares the obliquity of ecliptic to be 23 degrees only; and who, as if to show that it was no slip of the pen, assigns the longest day of 24 hours to the latitude of *sixty seven* degrees.

Now, the most remarkable anomaly attributable to Chaucer, is his adoption of 23° 50′ as the obliquity of the ecliptic; that being very much in excess of its known quantity in his time.

He says it is "after Ptholemy," which again is disputable on the ground that Ptolemy's obliquity was 23° 51′ 20″.

But M. Delambre, in his *Astronomie Ancienne* (Vol. I., pp. 87-88) shows, in an elaborate argument, that Ptolemy borrowed his obliquity from Eratosthenes, who had stated the ratio of the double obliquity to the circumference of the whole circle to be as 11 to 83; which ratio, worked out by Ptolemy *literally*, produced his obliquity above named. But, M. Delambre remarks, Eratosthenes had only intended, by 11 to 83, an approximate ratio to express an observed double obliquity of forty-seven degrees and two-thirds—sixths of degrees being the nearest approach his instruments were capable of. Now it is certainly remarkable that Chaucer's obliquity of 23° 50′ is *not* Ptolemy's but *is* the half of 47° 40′.

In an essay upon the meaning of Chaucer's *prime*, which will be found at the end of this volume, I produce another example of the general accuracy of his astronomical results, and deduce from it a conclusive argument in favour of an important correction of the text in "The Nonne's Preest's Tale."

Indeed, these researches into the astronomy of Chaucer are of great importance to the text of his poetical works.

In the first part of this Treatise (*infra* page 27) I have altered "eighte spere," of the copies, into ninthe speere, as applicable to "the firste mevyng of the firste mevable." Now, my justification for so doing happens to affect very materially an uncertain reading of a passage in The Frankelein's Tale; which again reciprocates the benefit by confirming in its context the propriety of the alteration alluded to: while at the same time it shows very conclusively what its own reading ought to be. The passage is as follows in Tyrwhitt's edition, and in all the editions previous to his :—

> And by his eighte speres in his werking
> He knew ful wel how fer Alnath was shove
> Fro the hed of thilke fix Aries above
> That in the ninthe spere considered is.
>
> *The Frankelein's Tale*, 11591.

But the two latest editors, Mr. Wright and Mr. Morris, both print "thre" and "fourthe," respectively, for *eighte* and *ninthe*. No doubt the alteration has the authority of the MS. selected by these editors as their original; but there cannot be a more convincing proof of its error than this diagram of the nine spheres, which is copied from a Treatise on the Sphere by Sacro Bosco, edited by Melancthon in 1538, but written by Sacro Bosco about a century and a half before Chaucer; and the diagram has every appearance of being a facsimile from the original MS.

FIGURA OSTENDENS
distributionem & ordinem
Sphærarum Cœlestium.

Here the Earth is in the centre, and around it are the nine spheres—viz., Moon, Mercury, Venus, Sun, Mars, Jupiter, Saturn, the Fixed Stars—denoted by the black ground with stars and the white band above it with the signs of the zodiac—and, lastly, the Ninth sphere, exterior to all.

The two last are very remarkable in being both distinguished by the signs of the zodiac, one by symbols and the other by effigies. And it must be observed that it is to the ninth sphere the effigies are given, a further proof, if any were necessary, that the animal configurations were attributed to the *true ecliptic*, and not to the stars which originally composed them. (See Note A in the Appendix.) Each of these zodiacs is divided in twelve equal parts; another proof that unequal divisions by constellations were not regarded in the middle ages.

The divisions in the ninth sphere diverge from those in the eighth by about one fifth of each sign, which very nearly agrees with the amount of precession fixed by Ptolemy, who made the longitude of the first star in his constellation Aries six degrees and a half from the equinoctial point.

Now this diagram of the spheres has a direct bearing upon the text of The Frankeleyn's Tale; not in the amount of the deviation, but in showing how the comparison between the two spheres might indicate to the astrologer "*how fer Alnath was shove.*" Alnath, being a fixed star, was in the eighth sphere, while "the hed of thilk fixe Aries above" was in the ninth sphere. Surely nothing could more plainly express this process by comparison of the two spheres than the passage in question:—

> And by his eighte spere in his werching
> He knewe ful wel how fer Alnath was shove
> Fro the heed of thilk fixe Aries above
> That in the ninthe spere considred is.

That is, he knew,—by the extent of the divergence between the divisions of the eighth and ninth spheres as shown in this diagram (or in some similar one in accordance with the precession at the time). For the ninth sphere was that of the equinoctial points, and as these were carried forward by the effect of precession in the same direction as the diurnal revolution, by so much was the ninth sphere supposed to exceed in swiftness the eighth and all the rest; dragging them forward, as it were, and causing them, in appearance, to strive back in a contrary direction:—

> "O firste mevyng cruel firmament,
> With thy diurnal swough that crowdest ay
> And hurlest al fro est to occident
> That naturally wold holde another way."
>
> *Man of Lawe's Tale*, 4715.

Therefore the readings *eighte* and *ninthe* in the Frankelein's Tale are clearly correct, while *thre* and *fourthe* convert the passage into unintelligible nonsense.

In another essay subjoined to this volume, wherein I endeavour to shew that by "the Carrenare" in the Booke of The Dutchesse, Chaucer meant the gulf of *Il Carnaro*, in the Adriatic sea, there will be found an extract from Sebastian Munster's Cosmographie, describing the intermittent lake near Zirck-nitz, in very nearly the same terms as a modern account of the same lake which appeared in the French periodical, "Cosmos," and was thence translated into "The Student," of September, 1869. That a phenomenon so singular and so apparently dependent upon accidental causes should be so little changed after the lapse of centuries, is extremely interesting.

I have also reprinted in this volume a series of papers, contributed by me to "Notes and Queries," in 1851, upon Chaucer's astronomy in the Canterbury Tales; a subject intimately germane to this Treatise on the Astrolabe. The few modifications in the views I then expressed which have occurred to me in a reperusal, I have explained in additional notes to those papers.

A. E. B.

Leeds, December, 1869.

EXPLANATION OF THE PLATES.

The lines and circles of the Astrolabe being stereographic projections on the plane of the equator, may either be drawn by construction, as practised and explained by Stoeffler, or calculated by trigonometrical formulae. The first method is rude and uncertain, depending upon ruled lines and their points of intersection, often at very acute angles. The second method is by far the more correct, being distances and radii calculated to any desired degree of exactness, and set off with precision from a scale of equal parts. It is by this second method that the following diagrams have been prepared.

PLATE I. Represents the front face of the outer or principal plate, serving as a frame to the other parts, and called by Chaucer "the Moder." It is "*thickest at the brynkes, that is, the utmost ryng with degrees; and al the myddel within the ryng shall be thynner to receyve the plates for diverse clymates, and also the rethe.*"

The hollow part thus formed is called by Chaucer "the wombe;" and the graduation of the outer circle of degrees, and of the inner hour-circle where the hours are denoted by the letters of the alphabet, is described in the text.

PLATE II, Represents the back or reverse side of the same plate. All the circles, scales, and divisions are marked upon it as described by Chaucer, with the exception of "the names of the holie daies in the Kalender and the letters A.B.C. on which thei fallen." These I have omitted for two reasons: first, because their selection and distribution must be purely arbitrary and conjectural, since Chaucer does not specify them. Second, because any such names and letters, without being of any real use to the illustration of the instrument, would have very much crowded and encumbered the drawing. Had it been necessary, however, to realize this part of the description, the most probable source for these names would be the old mnemonic lines known in mediæval times as the "*Cisio Janus.*"

It was on this reverse side of the plate that the radial index, with sights, revolved, for taking altitudes, celestial and terrestrial; and the following is the manner in which I imagine the index may have been centred so as to be independent of the movable centre-pin. A fixed solid boss may have been raised upon the plate, around the central hole, which, when truly turned, both interiorly and exteriorly, may have received the centre-pin in its interior and the index upon its exterior. Such a centring for an index stretching across the diameter of a circle, may be seen in the best sort of circular protractors, where the centre is perforated for the purpose of adjusting it to a point marked upon the paper beneath. In this way, and in no other that I can see, the centre-pin might be wedged against the rim of the boss without interfering with the free motion of the index.

It is not clear from Chaucer's description of "*the label*" whether it also revolved upon a centre, or whether it was detached, and used merely as a straight edge. I think the latter is the more probable; and that it was adjusted by the eye to a true diameter by its fiducial edge, which Chaucer calls its *point*. It is in this way I have represented it in Plates V. and VI., by a single straight line drawn diametrically across the centre. If, however, any person should prefer to believe that the label, like the index, revolved upon a centre, there can be no difficulty in imagining a prolongation of the head of the centre-pin, upon which both the rete and the label might turn. But Chaucer's description of the label is (page 31):—"Then thou hast a label that is shapen like a rule," (*i.e.*, like an index), "*save that it is straight*, and hath no plates" (sights) "on either ende; but with the small point of the forsaied label shalt thou," etc.

"*Save that it is straight*," can only mean that the fiducial edge was in one continuous straight line uninterrupted by any centring."

THE SCALE, as I have drawn it, requires especial remark. It is represented by Stoeffler, in depicting his own Astrolabe, and by MSS. 314 and 261 in depicting Chaucer's, as double; that is, extending on both sides, east and west of the vertical line. Now there is nothing in Chaucer's description to warrant this double extension. His words are (page 25):—

"Under the crosse line is markyd the scale, in maner of two squires, or els in manner of two ledders; that serveth by his XXII pointes, and by his devisions, of full many a subtil conclusion." In this description the XXII pointes are clearly the eleven steps or grades on each scale, numbered I to XI, which constitute its resemblance to a ladder: the XIIth is common to both scales and is not numbered. Were any further proof needed that the two MSS. above mentioned are sufficiently modern to have been copied in some particulars from Stoeffler, their representations of double scales in opposition to Chaucer's description would furnish it. Because, as I have remarked in a note to Chaucer's description, (*infra* page 25), the error of transposing umbra recta and umbra versa, putting each where the other ought to be, (which is common to all the printed copies and to MSS. 314 and 23002), also exists in Stoeffler's corresponding description in his own book: and I may add that in all probability it originated there, and was thence copied into edition 1532, and so was transmitted to every succeeding edition. For it is not at all likely that such an obvious error should originate independently in two different treatises on the same subject.

Stevins saw this error and avoided it in his own MS. and he also corrected it in the margin of MS. 314, by drawing his pen across *versa* and writing *recta* at the side. Whether the error really existed in MSS. anterior to Stoeffler's book can only be decided by comparison with some manuscript of *unquestionable* antiquity; in which, if the same error be present, then the only alternative will be to suppose that Stoeffler copied from some source common to both.

PLATE III.—The Rete. This is sufficiently explained in the text, p. 29. It is only necessary to add that the ribs which stretch from the margin and converge to the pole of the ecliptic, thus forming the net work of the rete, are so drawn as to be circles of longitude for the beginning of each of the twelve signs of the Zodiac.

PLATE IV.—" The circles under the rete" for the Latitude of Oxford. These circles are fully explained in the text, pp. 26—29.

PLATES V and VI.—The object of these two plates, which represent the two observations described by Chaucer in the IVth conclusion, pp. 33 to 35, is to place them in contrast for the purpose of demonstrating that the star which Chaucer calls Alhabor, in the second observation, was RIGEL, and not *Sirius* as has been hitherto dogmatically assumed.

First it may be seen, in plate V, that the sun is in a very favourable position for observation—that is, about mid-way between the horizon and the meridian on the eastern side; and, in plate VI, that the star Rigel occupies almost exactly the same position on the western side. This similarity of position must have been Chaucer's aim. Now Sirius, at the time stated, would be much nearer to the meridian—so near, indeed, as to closely border upon that proximity to it which Chaucer expressly condemns in this very conclusion. But, moreover, Sirius, at the stated time would not be at the required altitude of 18 degrees: and Stevins, in attempting to reconcile this difficulty, went so far as to alter the time of the observation from—"*passed seven of the clocke the space of eleven degrees*" (*i.e.* 44 minutes) into "*passed eight of the clocke the space of certaine minutes.*"

The second proof is derived from Chaucer's description of the star, which is, —"*amonge an hepe of stars* it liked me to take the altitude of the faire white starre that is cleped *Alhabor.*" A glance at the heavens, or at any celestial map will show at once the justness of this description as applied to Rigel, but its utter inapplicability to Sirius: there is no constellation more brilliantly furnished than Orion, nor a more solitary one than Canis Major.

And the third proof is derived from Fretagh's Arabic Lexicon, Vol. ii, p. 127. where may be found the Arabic name "Rijil al Habor. Nomen stellæ magnæ et lucidæ, β in pede Orionis."

After this, no further argument need be urged, I shall merely point out that in Plate VI, the position of Rigel is marked by the common intersection of its circles of longitude and latitude with the almicanter of 18 degrees.

In projecting these circles I have assumed for Rigel in the year 1390, a longitude of 68° 15′, and a latitude of 31° 23′ South.

Chaucer's Astrolabe.

Plate I.

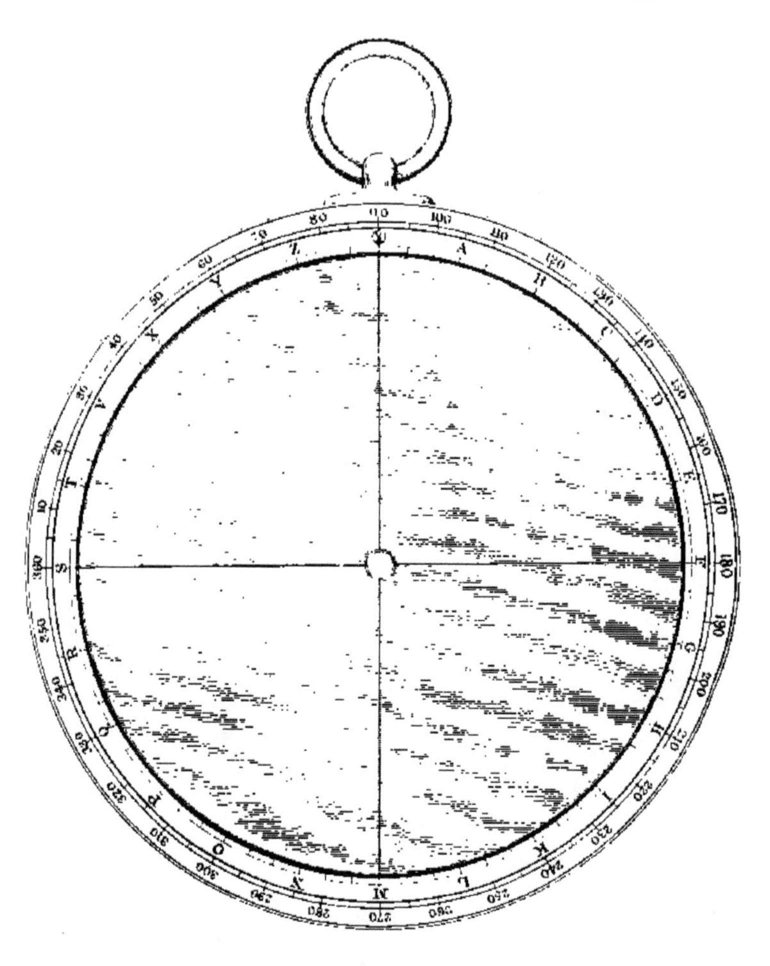

Litho Whiteman & Bass London

Chaucer's Astrolabe.
Plate II.

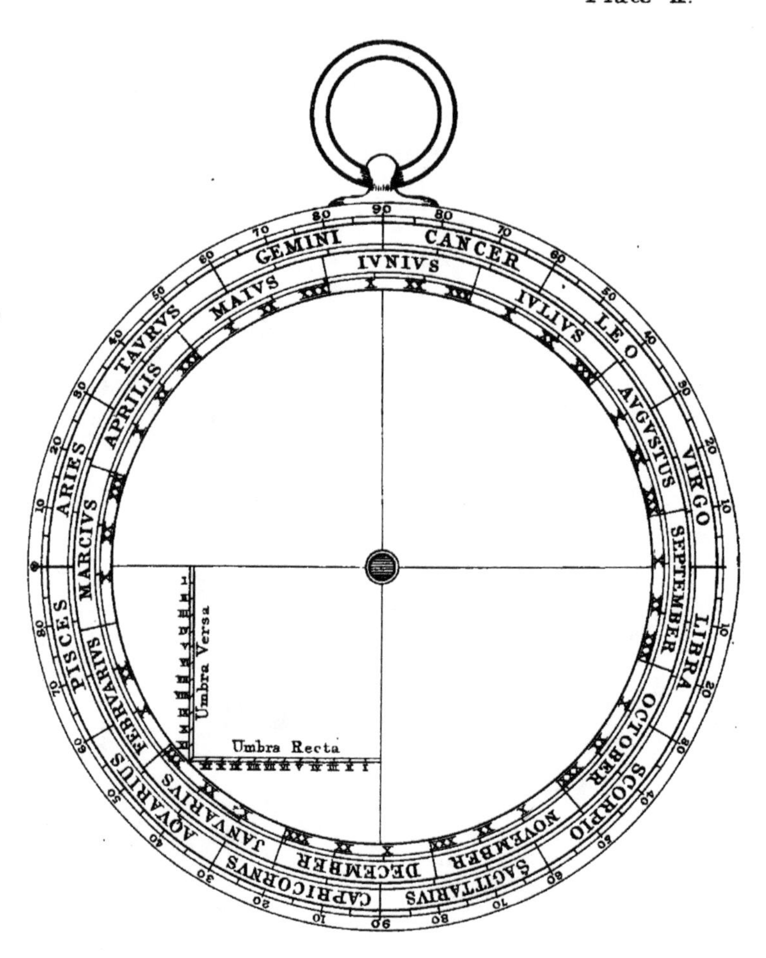

Lith. Whiteman & Bass, London

Chaucer's Astrolabe
Plate II

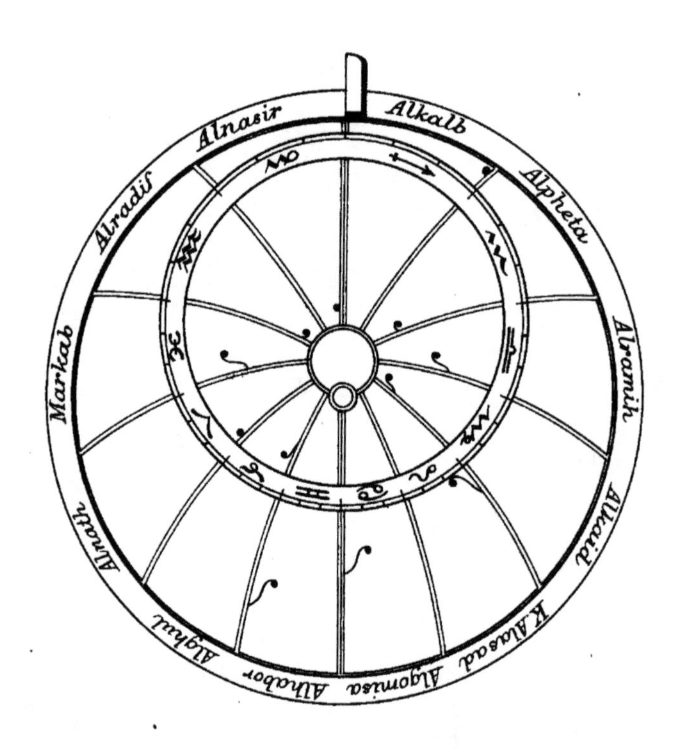

THE RETE OR ZODIAKE.

Alnasir — α *Lyræ*	*Alkalb* — α *Scorpii*
Alrudif — α *Cygni*	*Alpheta* — α *Cor. Borealis*
Markab — α *Pegusi*	*Alramih* — α *Bootis*
Alnath — α *Arietis*	*Alkaid* — η *Ursæ Majoris*
Alghul — β *Persei*	*K. Alasad* — α *Leonis*
'lhabor — β *Orionis*	*Algomisa* — α *Canis Minoris*.

Chaucer's Astrolabe

Plate I

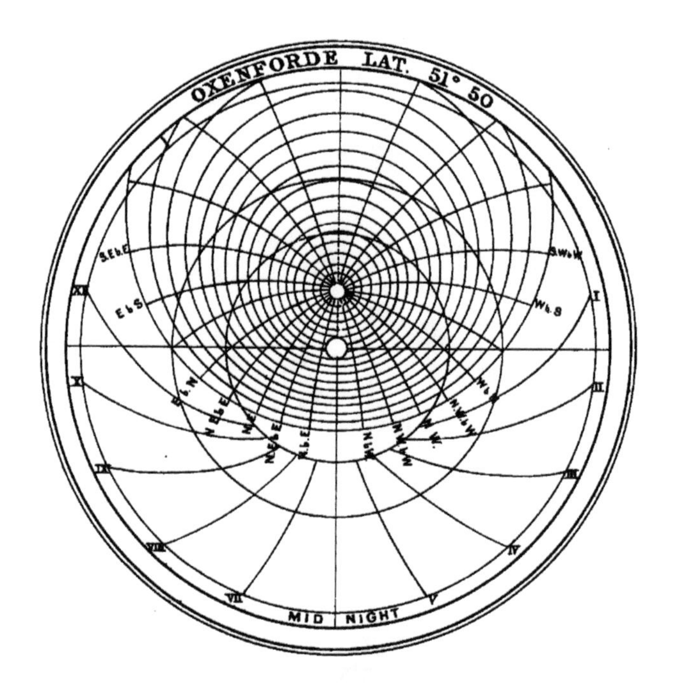

THE PLATE UNDER THE RETE.

Chaucer's Astrolabe

Plate V

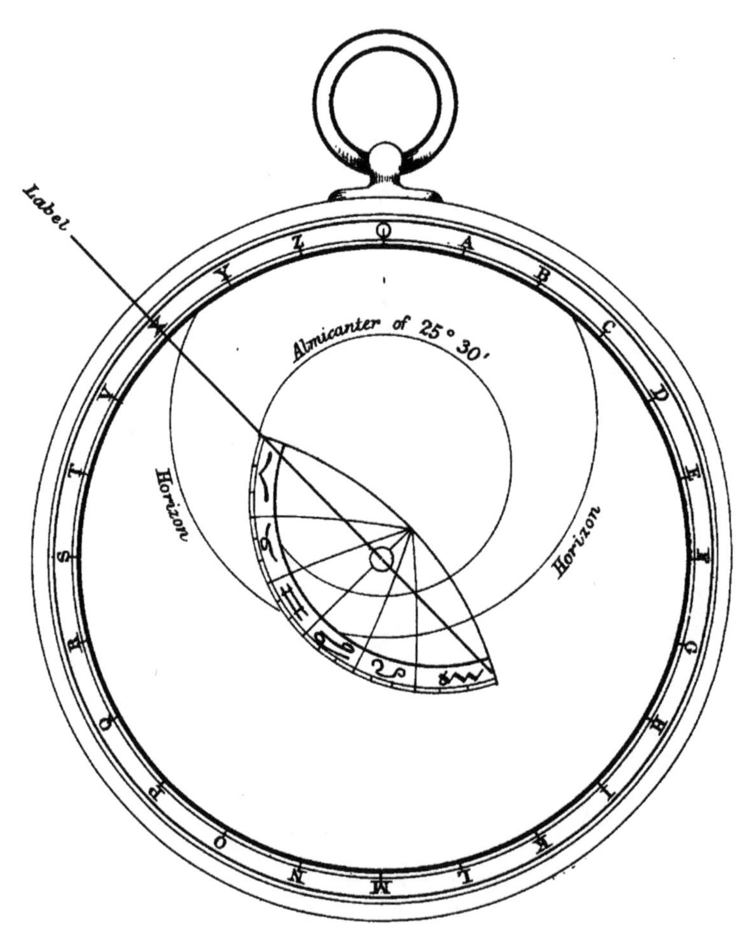

Diagram of Chaucer's First Ensample.

(of the Sun.)

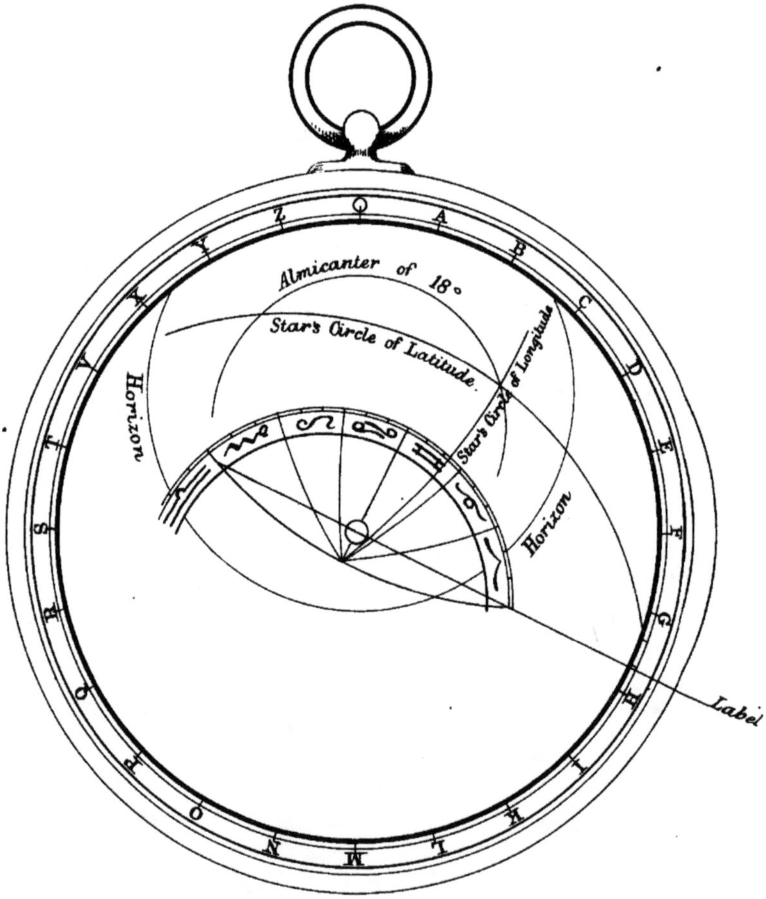

Chaucer's Astrolabe.

Plate V.

Diagram of Chaucer's Second Ensample.

(of the Star Rigil, called by him Alhabor.)

The Conclusions of the Astrolabie.

BY GEOFFRY CHAUCER.

THE PROHEME.

LYTEL LOWYS my sone, I perceyve wel by certeyn evydences thyn abylite to lerne sciences touchinge nombres and proporcions and also wel consydre I thy besye prayer in special to lerne the tretys of the Astrolabie. Then forasmoche as a philosopher sayeth he wrappeth hym in his frende that condiscendeth to the ryghtful prayers of his frende, therefore I have gyven thee a sufficient astrolabye for oure horizont compowned after the latytude of Oxenforde: Upon the which by medyacion of this lytel tretys I purpose to teche thee a certayn nombre of conclusions pertayning to this same instrument. I saye a certayn nombre of conclusions for three causes. The firste cause is this: Truste wel that alle the conclusions that have ben founden, or elles possiblye might be founde in so noble an instrument as is the astrolabye, ben unknowen perfitly to any mortal man in this region as I suppose: another cause is this, that sothely in any tretyse of the astrolabye that I have y-sene ther ben some conclusions that wol not in alle thinges perfourme ther behestes; and some of hem ben too harde to thy tender age of ten yere to conceyve.

This tretyse, devyded in five partes, wil I shewe thee undyr lighte reules and nakyd wordes in englissh—for latyn ne canste thou not but smal, my litel sone. But nevertheless suffyseth to thee these trewe conclusions in englissh as wel as suffyseth to the noble clerkys grekes these same conclusions in greke; and to the arabiens in arabike, and to jewes in hebrew, and to latyn folk in latyn: which latyn folk hadde hem first out of other divers langages, and wrote hem in ther own tonge, that is to saien, in latyn. And God wote that in alle these langages and in many mo have these conclusions ben sufficiently lernéd and taughte and yit by divers reules, right as dyvers pathes leden dyvers folk the right weye to Rome.

Now wol I pray mekely every discrete person that redeth or hereth this lytel tretyse to have my rude endyting excused, and my superfluite of wordes, for two causes: the first cause is, for that curious endyting and harde sentens is ful hevy at onys for suche a childe to lerne: and the second cause is this, that sothely me semeth better to writen unto a child twise a gode sentens than he forlete it ones. And Lowis if it be so that I shewe thee in my lyth englysh as trewe conclusions touchinge this matyr, and not only as trewe but as manye and as subtile conclusions as ben yshewed in latyn in any comune tretise of the astrolabye conne me the more thanke and praye God save the Kyng that is lord of this langage, and alle that him fayth bereth and obeyeth, everich in his degre the more and the lasse. But consydereth wel that I ne usurpe not to have founden this werk of my labour or of myn engyn: I n'am but a lewd compilatour of the labour of olde astrologiens, and have it translated in myn englysh only for thy doctryne: and with this swerd shal I sleen envye.

The firste partye of this tretyse shal shewe the figures and the membres of thyn astrolabye, bicause that thou shalte have the greter knowinge of thyne owne instrument.

The seconde partye shal teche thee to werken the verray practike of the forsayde conclusions as ferforth and also narrow as may be shewed in so smale an instrument portatife aboute. For wel wote every astrologien that smallest fractions ne wol not be shewed in so smale an instrument as in subtile tables calculed for a cause.

The thridde partye shal conteyne dyvers tables of longitudes and latitudes of sterres fyxe in the astrolabye. And tables of the declinacions of the sonne. And tables of the longitudes of cities and tounes. And tables as wel for the governaunce of the clokke as for to finde the altitude meridiane, and many an other notable conclusion after the kalenders of the reverente clerkes frere John Somme and frere N. Lenne.*

THE FOURTHE PARTYE shal be a theorike to declare the meving of the celestiale bodyes, with the causes; the whiche fourthe partye in special shal shewe in a table the verray mevyng of the mone from one to one, every daye, and every signe after thyn almanak. Upon the whiche table ther followeth a canon sufficient to teche as wel in maner of working in the same conclusions as to knowe in oure orizonte with whiche degre of the Zodiake the Mone aryseth in eny latitude, and the arysing of eny planete after his latitude fro the ecliptike lyne.

THE FYFTE PARTYE shal ben an introductorie after the statutes of oure doctours, on which thou maiest lerne a grete parte of the generalle rules of theorike in astrologye. In which fifte partie thou shalt finde tables of equacions of houses after the latitude of Oxenford; and tables of dignitees of plànettes and othere notefulle thinges if God vouchesauf and hys mothir the mayde, mo than I behote.

* Nicolaus de Lynna, *i.e.*, of Lynn, in Norfolk, was a noted astrologer in the reign of Edward III., and was himself a writer of a treatise on the Astrolabe. See Bale—who mentions "Joannes Sombe" as the collaborateur of Nicolaus—"Istos ob eorum eruditionem multiplicem, non vulgaribus in suo Astrolabio celebrat laudibus Galfridus Chaucer pœta lepidissimus."—BALE (edit. 1548), p. 152.

Here Endeth the Proheme.

THE FIRSTE PARTYE

Sheweth the Figures and the Membres of thyn Astrolabye.

The Ryng.

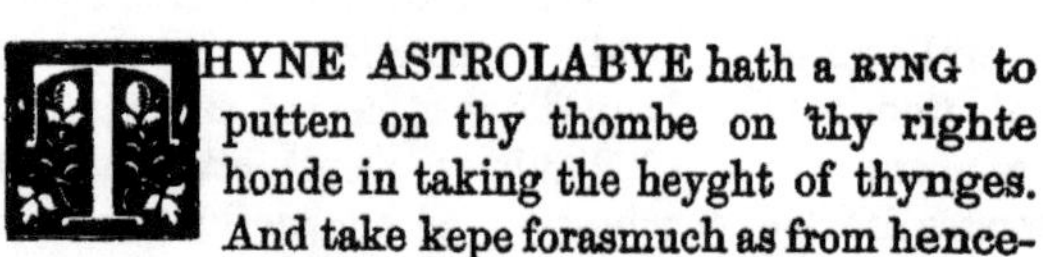

THYNE ASTROLABYE hath a RYNG to putten on thy thombe on thy righte honde in taking the heyght of thynges. And take kepe forasmuch as from henceforward I wol clepe the heyght of every thing that is take by the reule *the altitude* withouten mo words. This ryng renneth in a maner of a toret set fast in the mothir of thyn Astrolabye in a rume, or a space, that it distourbeth not the instrument to hangen after the right centure.

The Moder.

The moder of thyn astrolabie* is thickest by the brinkes, that is, the utmost ryng with degrees; and all the myddle within the ryng shall be thynner to receve the plates for divers clymates, and also for the rethe that is shape in manner of a net or elles after the webbe of a loppe.

This moder is devyded on the bakkhalfe with a lyne that cometh discendyng fro' the ryng down to the netherest bordure, the which lyne fro the forsaied ryng unto the centre of the large hole amidde, is cleped the South lyne or elles the lyne merydyonal: and the remenaunt of this

---

* This description of "the moder" is twice repeated in all the copies; the first time as an addition to the preceding description of the ring with which it has properly no connexion, and from which I have expunged it, retaining only this separate paragraph.

lyne doune to the bordure is cleped the North lyne or elles the lyne of mydnyght.

The Cross Lines.

Overthwarte this forsaied long lyne there crosseth hym an other lyne of the same length fro est to west: of the which lyne from a lytel crosse in the bordure unto the centre of the large hole is cleped the est line or elles the line orientale: and the remenaunt of the line fro the foresaied *centre** unto the bordure is y-cleped the west line, or the line occidentale.

The Foure Quarters of the Firmament.

Now hast thou heer the foure quarters of thyn astrolabye devyded after the foure principale plages or quarters of the firmament.

The Righte Side and the Lyfte.

The est-side of thyn astrolabie is cleped the righte side, and the west side is cleped the lyfte side. Forgete not this, litel Lowys. Put the ring of thyn astrolabye upon the thombe of thy righte hande and then wil his right side be toward thy lyfte side and his lyft side wil be toward thy righte side: and take this reule generale as wel on the bakk as on the wombe syde. Upon the ende of this est line, as I first saide, is ymarked a litel cros, whereas evermore generally is consydred the entryng of the est degre in the which the sonne aryseth.

The Degrees and the nombres of Augrime.

Fro the litel cros up to the ende of the meridional line, undyr the ryng, shalte thou finde the bordure devyded with xc degrees and by that same proporcion is every quarter of thyn astrolabye devyded: over the whiche degrees ther been nombres of augrim that devyden thilke same degrees from five to five, as sheweth by longe strikes betwene;† [of whiche longe strikes the space betwene conteyneth a mile-waie: and every degre of thilke bordure conteyneth foure minutes, that is to saie, foure minutes of an hour.]

* I have here substituted "*centre*" for "oriental" of the copies.

† Nombres of augrim, *i.e.*, Arabic figures. Stevens omitted the words in brackets—I think correctly; for they properly apply only to the other face of the instrument.

The names and the Signes.

Under the compase of thilke degrees ben wretyn the names of the twelve signes, as Aries, Taurus, Gemini, Cancer, Leo, Virgo, Libra, Scorpio, Sagittarius, Capricornus, Aquarius, Pisces. And the nombres of the degrees of these signes ben wretyn in augrim above, and with longe devysions from five to five devyded, from the tyme that the signe entreth, unto the last ende. But understande wel that these degrees of signes ben everych of hem consydred of lx minutes, and every mynute of lx secondes, and so forth unto smale fractions infynite, as sayeth Alcabucius.* And knowe wel that a degre of the bordure conteineth foure minutes; and a degre of a signe conteineth lx minutes: and have this in mind.

The Cercle of the Dayes.

Next this, followeth the cercle of the daies, figured in maner of the degrees, that containen in nombre thre hundred threescore and five: devyded also with longe strikes from five to five, and the numbres of augrim wretyn under the cercle.

The namys of the Monethes.

Next the cercle of daies, followeth the cercle of the twelve names of the monethes; that is to saie, Januarius, Februarius, Marcius, Aprill, Maius, Junius, Julius, August, September, October, November, December. These monethes taken ther names, some for properties, and some by statutes of Emperours, and some by other Lordes of Rome. Eke of these monethes, as liked to Julius Cæsar and Cæsar Augustus, some were compowned of divers nombres of daies as July and August. Then hath Januarius xxxi daies, Februarius xxviii, Marcius xxxi, April xxx, Maie xxxi, Junius xxx, Julius xxxi, August xxxi, September xxx, October xxxi, November xxx, December

* Et unumquidque istorum signorum dividitur in 30 partes equales quæ [gr]adus vocantur. Et gradus dividitur in 60 minuta; et minutum in 60 [secu]nda; et secundum in 60 tertia; similiterque sequuntur quarta; similiter [qu]inta; ascendendo usque ad infinita." Alchabitii, Diffᵗⁱᵃ 1ᵐᵃ.

xxxi.* Nathelesse although that Julius Cæsar toke two daies out of Feverire and put hem in his moneth of July; and Augustus Cæsar cleped the moneth of Auguste after hys name, and ordeined it of one and thirtie daies: yet truste wel that the Sonne dwelleth therefore never-the-more ne-the-lesse in one sygne than in an other.

The Holy-dayes.

Then foloweth the names of the holy-daies in the Kalender and next hem the letters of the A.B.C. on which thei fallen.

The Scale.

Next the forsaid cercle of the A.B.C., undyr the overthwarte lyne, is markyd the SCALE, in maner of two squires or elles in manner of two ledders, that serveth by hys twenty two poinctes, and hys devisions, of ful many a subtill conclusion. Of this forsaied scale, fro the crosse lyne unto the verie angle is cleped UMBRA VERSA and the nethyr partie UMBRA RECTA, or elles UMBRA EXTENSA.†

The Reule.

Then haste thou a brode reule that hath on every ende a square plate percyd with certayne holes, some more and some lasse, to receyve the stremes of the sonne by day, and eke by mediacion of thyn eye to knowe the altitude of the sterres by night.

The Pyn.

Then is ther a large pyn, in maner of an exiltre that goth thrugh the hole amyd, that halte the tables of the clymates in the rete in the wombe of the moder; throw which pyn ther goth a litel

* It is worthy of remark that Stevins here inserts in his MS., by way of illustrating the text, those well-known lines, "Thirtie dais hath September," &c., adding, "Lo, verses of the nomber of the dais in the usuall moneths of ye kalendar." This is, I think, an earlier date than has hitherto been discovered for these lines.

† In all the copies, printed and MS., except that of Stevins, these terms are reversed: except also Sloane MS. 314, in which the same error, *which originally existed in it also*, has been corrected in the margin, probably by Stevins. And it is very remarkable that Stœffler, who perhaps never saw Chaucer's treatise should yet, in *his* description of this part of the Astrolabie, have precisely th same error of reversal of terms in both his text and his illustrative drawing.

wedge the which is cleped the hors that streyneth alle these partes togeder. This forsaied gret pyn, in maner of an exiltre, is imagined to be the pole artike in thyn Astrolabye.

The Foure Quarters of the Wombe Syde.

The wombe side of thyn Astrolabye also is devided with a longe cross in foure quarters fro the est to west, and fro the south to north, from righte side to lyfte side, as is the bakksyde. The bordure of whiche wombe side is devided fro the poynt of the est line unto the poynt of the south line undyr the ryng in xc degrees; and by the same proporcion is every quarter devided as is the bakkside that amounteth thre hundred sixtie degrees. And understande wel that the degrees of this bordure ben answerynge and consentynge to the degrees of the equinoctial that is devided in the same nombre as every other cercle in the highe hevene.

The Degrees of the Bordure.

The xxiv. Houres.

This bordure is devided also with xxxiii letters and a smale cros aboute the south line that sheweth the xxiv houres equale of the clokke. And I have seide five of these degrees maken a mile-waie, and thre mile-wai maken an hour: and every degre of this bordure conteyneth foure minutes, and every minute foure* secondes. Now have I tolde thee twise for the more declaracion.

The Plate Under the Rethe.

The plate under the Rete is descryved with thre cercles, of whiche the leest is cleped the cercle of Cancer; bicause that the hed of Cancer tourneth evermore concentrikę uppon the same cercle. In this hed of Cancer is the gretest declynacion northward of the sonne, and therefore is he cleped Solsticium of Somer: which declynacyon, after Ptholemy is xxiii degrees and fyfty minutes, as wel in Cancer as in Capricorne. This cercle of Cancer is cleped the tropike of somer, of *tropos*,

The Solsticium of Somer.

---

* In editt. 1602, 1687, and in Urry's, this is misprinted "*fowertie secondes.*" In the earlier printed editt., and also in Stevins' and other MSS. it is "*lx secondes.*" See page 4 *supra* in Introduction.

that is to seyn, ayenward; for then begynneth the sonne to passe fro usward.

The myddyl cercle in widenesse of these thre is cleped the cercle Equinoctiale, uppon which tourneth evermore the heddes of Aries and Libra. And understonde wel that evermore thys cercle equinoctial tourneth justlie fro the verray est to the verray west, as I have shewed in the speere solide.

This same cercle is cleped also the Equatour or wayer of the day; for whan the sonne is in the hede of Aries or Libra than ben the dayes and nightes y-like of length in al the world, and therefore ben these two signes called equinoctis.

The Equatour.

And al that meveth within these heddes of Aries and Libra ben ycalled northward, and al that mevith without these heddes his movyng is cleped southward, as fro the equinoctial. Take kepe of this latitude north and south and forgete it not.

By this cercle equinoctial ben considered the xxiv houres of the clokke—for evermore the arising of xv degrees of the equinoctial maketh an hour equale of the clokke.

The xxiv Houres of the Clokke.

This Equinoctial is cleped the midway of the firste mevyng or elles of the sonne. Also it is cleped girdel of the firste mevyng for it departeth the firste movable in two like partyes even distaunt fro the poles of this world. And note that the firste mevygn of the firste movable is cleped mevyng of the [ninthe] speere, which mevying is fro est to west and ageyn into est.*

The widest of these thre cercles principale is cleped the cercle of Capricorn [bicause that the

* I have endeavoured by transposing some of the parts of this paragraph, and by changing " eighte" (sphere) of the copies into *ninthe*, to restore the sense. My reasons for this last alteration are given in the Introduction (*supra* p. 14.) The "firste movyng of the first mevable" which sounds so tautological may be explained by the two movings, first and second, into which Ptolemy divided the *Primum Mobile*. See the marginal note to The Man of Lawe's Tale 4715 in the Lansdown MS. reprinted by the Percy Society, edited by Mr. Wright, Vol. i, page 213.

head of Capricorne]* tourneth evermore concentrike uppon the same cercle. In the hede of this foresaied Capricorne is the gretest declinacion southward of the sonne and therefore it is cleped Solsticium of winter. This cercle of Capricorne is also cleped the tropike of winter, for thanne begynneth the sonne to come ageyn to usward.

The Solsticium of Winter.

The Almicanteras.

Uppon this forsaied plate ben compassed certeyne cercles that highten Almicanteras: of whiche some of hem semen perfite cercles and some semen imperfite. The centure that standeth amydst the narrowest cercle is cleped the Signet.† And the netherest cercle [or the firste cercle, is cleped the orizonte, that is the cercle]‡ that devideth the two emisperies, the partye of the heven above the yerthe and the partye beneth. These almicanteras be compouned by two and two, al be it so that on diveres astrolabyes some almicanteras ben devided by one, and some by two, and some by thre, after the quantite of the astrolabye. This foresaied signet is y-magined to be the verray poynt over the crowne of thy heed, and also this signet is the verray pole of the orizonte in every region.

The Orizont.

The Signet.

From this signet, as it semeth, ther comen croked strikes, like to the clawes of a loppe, or elles like to the worke of a woman's calle, inkerving overthwart the almicanteras; and these same strikes or devisions ben cleped Azimutes: § and they deviden the orizonte on thyne astrolabye in XXIV devisions. And these azimutes serve to knowe the costes of the firmament; and to other

The Azimutes.

---

* These words are not in the copies. I have taken them from the corresponding description of Cancer, just before; which Stevins also adopted.

† Stevins invariably but very improperly, altered *signet* to *Zenith* See notes to conclusions XVI and XXXII.)

‡ The words in brackets are not in the printed copies: they are supplied [illegible] the MSS.

§ For these azimuths, see Note to Conclusion XXXIII. page 52.

conclusions—as for to knowe the signet* of the sonne and of every sterre.

The Planetary Houres.

Next these azimutes, undyr the cercle of Cancer, ben the twelve devisions embolite, moche like to the shape of the azimutes, that shewen the spaces of the houres of the Planets.

The Rethe.

The Rete of thyn astrolabye, which is thy Zodiake, shapen in maner of a net, or of a lop webbe, after the olde descripcion, which thou maiest tourne up and doune as thiself liketh, conteynth certayn nombre of sterres fixe, with ther longitudes and latitudes determinate, if it so be that the maker have not erred. The namys of the sterres ben wretyn in the margin of thy rete, there as they sitte; of the whiche sterres, the smale poynt is cleped the centure. And understonde that alle the sterres sittynge within the Zodiake of thyn astrolabie ben cleped sterres of the north, for they arisen by the northest lyne: and al the remenaunt fixed out of the Zodiake ben i-cleped sterres of the south,—but I saie not that they arisen alle by the south-est lyne, witnesse of Aldebaron and also Algomisa. Generally understonde this reule, that thilke sterres that ben cleped sterres of the north arysen rather then the degre of ther longitude, and all the sterres of the south arysen aftyr the degre of ther longitude, that is to saien, sterres in thyn astrolabie.

The Starres Fixe.

The mesure of longitude of sterres is y-taken in the lyne ecliptike of heven; undyr the whiche lyne when the sonne and the mone ben lyne right, elles in the superficie of this lyne, thanne is the eclips of the sonne or of the mone: as I shall declare and eek the cause why. But sothelie the ecliptike lyne of the zodiake is the utterest bordure of the Zodiacke there thy degrees ben markyd. The zodiake of thyn astrolabie is shapen as a

* Here "Signet" means the azimuthal point, or bearing.

compace which that conteyneth a large brede as after the quantite of thyn astrolabye, in ensample that the zodiake of hevene is ymagined to be a superficies conteyning the latitude of twelve degrees whereas al the remenaunt of cercles in heven ben ymagined verray lines withouten any latitude.

The Ecliptike Lyne. Amyddes the celestial zodiake is imagined a lyne which that is cleped the ecliptike lyne, under the whiche lyne is evermore the way of the sonne. Thus ben there sixe degrees of the zodiake on that one side of the lyne, and six degrees on that other.*

The Zodiake. The zodiake is devyded in twelve principale devisions that departen the twelve signes: and, for the streightnes of thyn astrolabye, thanne is every smale devision in a signe y-parted by two degrees and two, I mene degrees conteyning lx minutes. And this hevenish zodiake is cleped the cercle of the signes or the cercle of bestes; for zodiake in language of greeke souneth bestys in latyne tonge. And in the zodiake ben the twelf signes that have names of bestes: bicause whan the sonne entreth in eny of the signes he taketh the propertie of soche bestes: or elles for that the sterres that there ben fixe ben disposed in signe of bestes, or in shape like bestes,† or elles when planetes ben under the signes thei transmue us by ther influence, operacions, and effectes like to operacions of bestys. And understonde also that when an hote planet cometh into an hote signe thanne encreaseth his hete, and if a planet be colde thanne amenuseth his coldnes bicause of the hote signe: and by this conclusion maiste thou taken ensample in alle signes be thei moiste

* There is some confusion here between the *celestial* zodiac and the instrumental zodiac of the astrolabe; six degrees on *each* side of the ecliptic cannot [illegible] the latter, for Chaucer had, a few lines before, described the ecliptic [illegible] the instrument as "the utterest bordure of the zodiake."

† Note B, in appendix, page 84.

or drie, movable or fixe, rekening the qualite of the planetes as I first saied.

The Label.

Thanne haste thou a Label that is shapen like a reule save that it is streight and hath no plates on eythere ende: but with the smale poynt of the foresaied label shalte thou calcule the equacions in the bordure of thyn astrolabye as by thyn almurie.

The Almurie or Denticle of Capricorn.

Thyn Almurie is cleped the denticle of Capricorn or elles the Calculere. This same almurie is set fixe in the hed of Capricorn, and it serveth of many a necessarie conclusion in equacion of thynges, as shal be shewed.

Here Endeth the Firste Partye.

THE SECONDE PARTYE.

Techeth the Pratike and the Conclusions of thyn Astrolabye.

I. To finde the degre in the which the sonne is daie by daie aftyr his course about.

RECKON and knowe which is the daie of the moneth and laie thy rewle upon the same daie, and then will the verrey poynt* of thy rewle sitten in the bordure upon the degre of the sonne. Ensample as thus: the yere of oure Lord a thousand thre hundred ninetie and one, the xii daie of March, at middaie, I would knowe the degre of the sonne.

I sought in the bakke halfe of myne astrolabie and founde the cercle of the daies the which I knewe by the namys of the moneths wrytten undyr the same cercle. Tho laied I my rewle over the foresaied daie and founde the point of my rewle in the bordure upon the first degre of Aries, a littel within the degre: and thus knewe I this conclusion. An other daie I would knowe the degre of my sonne, and this was at middaie in the xiii daie of December, I founde the daie of the moneth in maner as I saied: tho laied I my rewle upon the foresaied xiii daie and founde the poinet of my rewle upon the first degre of Capricorne a littel within the degre: then hadde I of this conclusion the full experience.

---

e and elsewhere, by the *point*, or small point, of the Rule, Chaucer r means its fiducial edge: and the same remark applies to the

II. To knowe the altitude of the sonne, eyther of celestiale bodies.

Sette the ryng of thyne astrolabie upon thy ryghte thombe and tourne thy lyfte syde again the light of the sonne and remeve thy rewle up and downe till the streme of the sonne shyne through bothe holes of the rewle: loke then how many degrees this rewle is areised fro the littel crosse upon the Est lyne and take there the altitude of thy sonne. And in this same wise maiest thou knowe by nighte the altitude of the mone or of the brighte sterres.

This chapiter is so general ever in one that there nedeth no more declaracion; but forget it not.

III. To knowe [by] the degre of the sonne, and of thy zodiake the daies in the backsyde of thyne Astrolabie.*

Then if thou wilte wete the reckenyng to knowe whiche is the daie, in thy kalender, of the moneth that thou art in, laie [the rule of] thyne astrolabye, that is to saien the alidatha, upon *the degre of sonne* and he shall shew thee *the daie of the monethe* in thy kalender.

IV. To knowe every tyme of the daie, by light of the sonne, and every tyme of the night by the sterres fixe; and eke to knowe by night or by daie the degre of the signe that ascendeth on th' est orizonte which is cleped comenly the ascendent.

Take the altitude of the sonne when thou liste, as I have said, and sette the degre of the sonne, in case that it be before the middel of the daie, among thyn almicanteras on the est syde of thyne astrolabie: and if it be after the middel of the daie sette the degre of the sonne upon the west syde. Take this maner of settyng for a generall reule, ones for ever.

And when thou haste isette the degre of the sonne upon as many almicanteras of height, as was the sonne taken by thy reule,

---

* This third conclusion was evidently meant to be converse to the first conclusion. The object of the first was to find the place of the sun in the ecliptic for any given day and month—so the object of this should be to find, conversely, the month and day for any given place of the sun. This intention is plain from the first two lines which remain unchanged. But in all the copies printed and MS. that I have examined, this sense is reversed by a stupid and incongruous confusion of terms, by which the proposition is made to appe repetition of the first conclusion instead of its converse. I have endeav to restore the original intention by altering in the title the place of the w brackets and by retransposing the terms at the end as indicated by i Stevins omits this conclusion altogether.

laie over thy label upon the degre of the sonne and then woll the point of the label sitten in the bordure upon the very tide of the daie. Ensample as thus: The yere of oure Lorde a thousand thre hundred ninetie and one, the twelveth daie of March, I would knowe the tide of the daie. I toke the altitude of my sonne and founde that it was xxv degrees and xxx minutes of height of the bordure in the bakksyde: tho tourned I myne astrolabie, and bicause it was before middaie I tourned my rete and sette the degre of the sonne, that is to saie the firste degre of Aries, in the righte side in myne astrolabie, upon the xxv degre and xxx minutys of height emong my almicanteras. Tho laied I my label upon the degre of my sonne and founde the point of my label in the bordure on the capital letter that is cleped an X: tho reckened I all the capitale letters fro the lyne of midnight unto the foresaied letter X, and founde it was nine of the clocke of the daie. Tho loked I over my Est orizont and found there the twentieth degre of Geminus* ascending, which that I toke for myne ascendente. And in this wise hadde I the experience for evermore in which maner I should knowe the tide of the daie, and eke myne ascendent.

Tho would I wete, that same night following, the hour of the night, and wrought in this wise: Among an hepe of sterres fixe it liked me to take the altitude of the faire white sterre that is cleped ALHABOR, and founde her sytting on the west side of the lyne of middaie, eightene degrees of heyth taken by my reule on the bakksyde.

Then sette I the centure of this Alhabor upon eightene degrees emong my almicanteras, upon the west side, bicause that she was found upon the west side. Tho laied I my label over the degre of the sonne, that was discended under the west orizont, and rekened alle the letters capitales fro the lyne of middaye unto the poynt of my label in the bordure and founde that it was after none, passed seven of the clokke the space of eleven degrees.† Then

---

* "*Twelveth* degree of Geminus" in all the printed copies; but, in the MSS., "the *twentieth* degree;" which last is correct. See the diagram of the position of the astrolabe in this observation, in Plate V.

† "Eleven degrees"; that is, forty-four minutes in time. Stevins, in his MS., very unwarrantably altered this time to—"past eight of the clocke certaine minutes"—his reason for doing so being obviously to reconcile the observation with the position of the star *Sirius*, then and since erroneously assumed to be Chaucer's star Alhabor. But I have asserted and proved that Chaucer's Alhabor, in this conclusion, was no other than the star Rijel (β Orionis). See Plate VI., and its explanation, *supra* page 18.

loked I doune uppon my est orizont and founde there twenty* degrees of Libra ascending whom I toke for myn ascendant: and thus lernede I ones for ever to knowe in which maner I should come to the hour of the night and to myne ascendent as nerely as maie be taken by so smal an instrument.

But natheles this reule in general wil I warne thee of for ever,— ne make thou never non ascendent at none of the daye: [ne canste thou nat] take a juste ascendent of thyne astrolabie and have sette justlie a clokk when any celestial body, by which thou wenest governe thilke thynges, ben neigh the southe lyne: for truste wel when the sonne is nere the meridionale lyne the degre of the sonne renyth so long concentrike upon thyne almicanteras that sothly thou shalte erre fro the juste ascendent. The same conclusion saie I by my centure of my sterre fixe by the nighte: and moreover by experience I wote wel that for oure orizont fro eleven of the clokke unto one, in taking the juste ascendent in a portatife astrolabye, it is to harde to knowe. I mene fro eleven of the clokke before none til one of the clokke next following.†

[All the printed editions, of which there were no less than eight, from 1532 to 1721, insert a conclusion in this place which is a plain interpolation, or rather a repetition almost word for word of a preceding conclusion (No. III). I have not found it in any MS., and its re-appearance in edition after edition of the printed copies is a curious exposure of the want of care and discrimination with which these editions were blindly copied one from another, notwithstanding the boast in some of them of having been "compared with many valuable MSS."]

4

V. Special declaracion of the Ascendent.

The ascendent, sothelie, as well in all nativities, as in questions, and as in eleccions of tymes, is a thyng which that these astrologiens gretlie observen. Wherefore me semeth convenient, sens I speke of the ascendent, to make of it a special declaracion. The ascendent, sothelie, to take it at the largest, is thilke degre that ascendeth at any of these foresaied tymes on the Est orizont: and therefore if that any planet ascende at thilk same time, in the foresaid same degre of his longitude, men saie that thilk planet is in horoscopo. But sothelie the hous of the ascendent, that is to saie

---

* *Eighteen* degrees of Libra would be more correct, and probably what Chaucer wrote.

† In the case of a star or planet, Chaucer of course means within fifteen degrees on either side of the meridian.

the firste hous, or the Est angle, is a thyng more brode and large: for, after the statutes of the Astrologyens, what celestial body that is v degrees above thilk degre that ascendeth on the orizont, or within that nombre, that is to saien, nere the degre that ascendeth, yet reckyn thei thilk planet in the ascendent: and what planet that is undyr thilk degre that ascendeth the space of xxv degrees yet saien thei that planet is like to hym that is the hour [lord ?] of the ascendent.*

But sothelie if he passe the boundes of the foresaide spaces above or beneth, thei saien that thilke planette is fallyng fro the ascendent: yet saien these Astrologiens that the ascendent, and eke the lord of the ascendent, maie be schapen for to be fortunate or infortunate, as thus: a fortunate ascendent clepen thei when that no wicked planet of Saturne or Mars, or els the Taile of the Dragon is in the hous of the ascendent, ne that no wicked planet have no aspecte of enmitie uppon the ascendent. But thei will caste that thei have fortunate planet in ther ascendent, and yet in his felicitie, and then saie thei that it is wel. Furthermore thei saine that fortune of an ascendent is the contrarie of these foresaide thinges. The lord of the ascendent saine thei that he is fortunate when he is in gode place for the ascendent; and eke the lord of the ascendent is in an angle, or or in a succedent, where he is in his dignitie and comforted with frendly aspectes receved; and eke that he maie seen the ascendent not retrograde ne combost, ne joined with no shrew in the same signe, ne that he be not in his discencion, ne reigned with no planet in his discencions, ne have uppon hym none aspect infortunate; and then thei saien that he is wel.

Nathelesse these ben observaunces of judicial matter and rites of Painims, in whiche my spirit hath no faith, ne knowyng of ther horoscopum: for thei saien that every signe is departed in thre evene partes by x degrees, and the ilke porcion thei clepen a Face: and although a planet have a latitude fro the Ecliptike yet saien

---

* Here Chaucer digresses into technical astrology: but since he expressly disclaims any knowledge of it, it is not necessary to discuss his meaning. The number of the degrees, however, of the ascendant above and beneath the horizon, being obviously taken from the Tetrabiblos of Ptolemy, the following numbers in the Latin translation will serve to justify the numbers in the text.

"*Quinque* gradibus qui super horizontem ante ipsum ascenderunt usque ad *viginti quinque* qui ad ascendendum remanserint." Lib. iii, (*De Loco Prorogatore.*) 12

All the printed copies and MSS. have xv instead of xxv for the last number: Stevins has 15 in both places.

some folk so that the planet arise in that same signe with any degre of the foresaied face in which his longitude is reckened, yet is that planet in horoscopo, be it in nativities or in eleccion.

5 VI. To knowe the verie equacion of the degrees of the sonne, if it so be that it falle betwixt two almicanteras.

For as moche as the almicanteras of thyn astrolabie ben compouned by two and two, whereas some almicanteras in sondrie astrolabies ben compouned by one, or elles by two, it is necessarie to thy lernyng to teche thee first to knowe and werke with thyn own instrument. Wherefore when that the degre of the sonne falleth betwene two almicanteras, or elles if thyne almicanteras ben graven with overgrete a poynt of a compace (for bothe these thinges maie cause errour as wel in knowing the tide of the daye as of the verray ascendent) thou must werken in this wyse: sette the degre of the sonne uppon the heigher almicantera of bothe and waite wel where the almurie toucheth the bordure and sette ther a prick of ynke; sette adoun agayn the degre of the sonne uppon the nethyr almicantera of bothe and sette there another pryck: remeve than thy almurie in the bordure even amiddes bothe pryckes and this wol leden justlie the degre of the sonne to sitte betwene bothe the almicanteras in his ryghte place.

Laye thanne thy label on the degre of the sonne and finde in the bordure the verray tyde of the daye or of the nyghte. And also verraily shalt thou finde upon thyn este orizont thyn ascendent.

6 VII. To knowe the spring of the dawning and the ende of the evenyng; the whiche ben cleped the two crepusculis.

Sette the nadyr of thy sonne upon eyghtene degrees of height amonge thyn almicanteras on the west side, and laye thy labell on the degre of the sonne; and then shall the point of thy labell shewe the spring of the daie: also sette the nadyr of the sonne upon the eyghtene degrees of heyght among thine almicanteras on the est side, and laye over thy labell upon the degre of the sonne, and with the point of thy label finde in the bordure the ende of thy evenyng, that is, verie night. The nadyr of the sonne is thilke degre that is opposyte to the degre of the sonne in the VIIth signe*, as thus: everye degre of Aries by order is nadyr

* For the extraordinary errors of the printed editions in this place, see the Introduction, page 7.

to everye degre of Libra by order; and Taurus to Scorpion; Gemini to Sagittarius; Cancer to Capricorn; Leo to Aquary; Virgo to Pisces. And if any degre in thy Zodiake be dark, hys nadyr shall declare hym.

VIII. To knowe the arche of the daye that some folk callen the daye artificialle, fro the sonne arysing tyl it goo downe.

Sette the degre of the sonne upon thyn est orizonte and laie thy label on the degree of the sonne and at the point of thy label in the bordure sette a pryck: turne than thy rete about tyl the degre of the sonne sytte upon the west orizonte and laie thy label upon the same degre of the sonne, and at the poinct of thy label sette another pryck. Reken than the quantyte of tyme in the bordure betwixe bothe pryckes and take there thyn arche of the daie. The remnaunt of the bordure undyr the orizonte is the arche of the nyghte. Thus maiest thou reken bothe arches of everye porcion where that thou likest: and by this maner of werkyng maiest thou se how longe that any starre fyxe dwelleth above the erthe fro the tyme that he riseth tyl he goo to reste. But the daie naturelle, that is to saien 24 houres, is the revolucion of the equinoctial with as moche partye of the zodiake as the sonne of hys proper mevynge passeth in the menewhile.

IX. To turn the houres inequalles into houres equalles.

Knowe the nombre of the degrees in the houres inequalles and departe hem by fiftene and take there thyne houres equalles.*

X. To knowe the quantyte of the daye vulgare, that is to say, fro the spryng of the daye unto the very nyght.

Knowe the quantyte of thyne crepusculis as I have it taught thee in the chapiter before, and adde hem to the arche of the daye

* This conclusion is very suspicious, and is probably an interpolation. If Chaucer himself had written it, he would not afterwards condemn it as superfluous: "The quantytes of houres equalles ben departed alredy in the bordure of thyn astrolabye — *what nedeth auy more declaracion.*"—See XIIth Conclusion. It is, moreover, out of place, since the reader knows nothing, *as yet*, of houres inequalles. Stevins shows that he saw this last objection by his having removed this conclusion, and placed it after those in which houres inequalles are explained. Another suspicious feature in this conclusion is its awkward intrusion in this place between the VIIIth and Xth, which ought to be consecutive, as required by Chaucer's words in the Xth:—"as I have taught thee in the chapiter before," meaning the VIIth, which should be changed so as to immediately precede the Xth.

artificiall, and take there the space of al the whole day vulgare unto the very nyght. In the same maner mayst thou werke to knowe the vulgare nyght [by abating the same crepusculis from the arche of the nyghte].*

10 XI. To knowe the houres inequalles by daye.

Understonde wel that these houres inequales ben cleped houres of the planettes: and understonde wel that sometyme ben they longer by daye than they be by nyght, and sometyme the contrarye. But understande thou wel that evermore generally the hour inequale of the daye with the hour inequale of the nyght conteineth 30 degrees of the bordure; the which bordure is evermore answeringe to the degrees of the equinoctiall. Wherefore departe the arche of the daye artificiall in 12 and take there the quantyte of the houre inequale by daye; and if thou abate the quantyte of the houre inequale out of 30† degrees then shall the remenaunt performe the houre inequale of the nyght.

11 XII. To knowe the quantyte of houres equalles.

The quantytes of houres equalles, that is to saien the houres of the clokke, ben departed by fyftene degrees alredy in the bordure of thyn astrolabye as wel by nighte as by daye generally for evermore.—what needeth any more declaracion. Wherefore whan thou liste to knowe how many houres of the clokke ben passed, or [how many]‡ of these houres ben to comen fro soche a tyme to soche a tyme by daye or by nighte, knowe the degre of thy sonne and laye thy label on it [and bring it to the este orizont and take there the tyme of the sonne arysing by thy label in the bordure],||

---

* Stevins saw the want which I have here supplied in brackets; but he remedied it by the addition of an entirely new conclusion as follows. "To know the quantity of the vulgar night. Take away both crepuscles from the arche of the night, and the vulgar night onlye remayneth." Now this interpolation of Stevins shows the probability of the IXth being also the work of some previous interpolator.

† 360 degrees in the printed copies. So also Stevins, only that he puts "houres" in the plural, which apparently lessens the error.

‡ Substituted for "any parte of any" of the printed copies.

|| I have inserted this clause (in brackets) as necessary not only to the sense, but because there is no other conclusion that teaches the time of sun-rise: an omission the more remarkable because the point of the sun's arising is referred to in the very next sentence, and again in the following conclusions.

thanne turne thy rete aboute joyntlie with thy label and with the point of it recken in the bordure fro the sonne arysyng unto the same place there thou desirest by daye as by nyghte. This conclusion woll I declare in the fowerthe partye of the laste chapiter of this tretyse so openly that there shal lacke no word that nedeth declaracion.

12 XIII. Speciall declaracion of the houres of the planettes.

Understonde wel that evermore fro the arysying of the sonne tyl it go to rest, the nadyr of the sonne shall shewe the houre of the planet; and fro that tyme forward al the nyght tyl the sonne aryse; then shall the very degre of the sonne shewe the howre of the planet. Ensample as thus: the 13 daye of March fel upon a Satyrday peraventure, and at the arysyng of the sonne I founde the seconde degre of Aries syttynge upon myn Est orizonte al be it was but lytel. Then founde I the seconde degre of Libra nadyre of my sonne, discendynge on my West orizonte, upon which West orizont every daye generally at the sonne arysyng entreth the hour of any planet undyr the foresaied West orizonte, after the whiche planete the daye bereth his name, and endeth in the nexte strike of the planete undyr the foresaied Weste orizonte: and ever as the sonne clymbeth upper and upper, so goth hys nadyre downer and downer, and echinge* fro such strikes the houres of the planettes by order, as they sytten in heven. The fyrst hour inequale of every satyrday is to Saturne, and the second to Jupiter, the thyrd to Mars, the fourth to the Sonne, the fyfth to Venus, the syxt to Mercurius, the seventh to the Mone, and then ayen the eyghth to Saturne, the nyneth to Jupiter, the tenth to Mars, the eleventh to the Sonne, the twelfth to Venus. And nowe is my sonne gon to reste as for that satyrdaye; Than sheweth the very degre of the sonne the hour of Mercury entryng under my west orizont at even: And nexte hym succedeth the Mone, and so forth by order, planete after planete, in hour after hour al the nyght longe tyl the sonne aryse. Nowe riseth the sonne the Sunday by the morowe, and the nadyr of the sonne upon the west orizont sheweth me the entryng of the hour of the foresaied sonne. And in this maner succedeth planete undyr planete fro Saturn unto the Mone,

---

* Stevins, not understanding this word echinge, changed it in his MS. to *teachinge:* but it has every appearance of being a genuine word, and the verb "to eche" is found in such a comparatively modern authority as Bailey's Dictionary Here it seems to mean adding one by one, or dealing out.

and fro the Mone up ageyn to Saturn, hour aftyr hour generally, and thus knew I this conclusion.

19 XIV. To knowe with which degre of the Zodiake any sterre fyxe, in thyn astrolabye, aryseth upon the Est orizonte, although the orizonte be in anothyr sygne.

Sette the centure of the sterre upon the Est orizonte and loke what degre of any signe that sytteth upon the same orizonte at the same tyme: and understonde wel that with the same degre ariseth the same sterre.

And this mervaylous arysing with a straunge degre in another sygne is bycause the latitude of the sterre fyxe is eyther north or south fro the Ecliptike.* For sothely the latitudes of planetes ben comenly rekenyd fro the ecliptike bycause that none of hem declineth but fewe degrees fro the brede of the Zodiake. And take gode kepe of thys chapiter of arysing of celestiall bodyes, for trusteth wel that neyther Mone, neyther sterre, in oure embolyfe orizonte that aryseth with the same degre of his longitude save in one case, and that is when thei have no latitude fro the eclyptike lyne. But nevertheles somtyme is everiche of these planetes undyr the same lyne.

20 XV. To knowe the declinacyon of any degre in the Zodiake fro the equinoccial cercle.

Sette the degre of any signe upon the lyne meridional and reken his altytude in the almicanteras fro the Est orizonte up to the same degre sette in the forsaied lyne, and sette there a prycke: turne up thy rete and sette the hed of Aries or Libra in that same merydyonal lyne and sette there another prycke; and when this is done consyder the altitudes of hem bothe: for sothely the difference of thilke altytude is the declynacyon of thilke degre fro the equinoctial. And if it so be that thilke degree be northward fro the equinoctiall than is hys declynacyon north, and if it be southward than it is south.

21 XVI. To knowe for what latytude in any region the almicanteras in thy tables ben compowned.

* In all MSS. and printed copies that I have examined this word is "Equinoctial," and the next word "But." These I have altered into *Ecliptike* and *For*, respectively: not so much because of the expression "latitude," which Chaucer often uses in the sense of declination; but because the problem necessarily requires the star's divergence from *the ecliptic*, while with the equinoctial it might *coincide*, and still be within the conditions of the problem.

Reken how many degrees of almicanteras, in the meridionall line, be from the circle equinoctiale unto the signet;* or els from the pole artike unto the north orizont: and for so grete a a latitude, or so smale a latitude is the table compouned.

13 XVII. To knowe the altytude of the sonne in the myddes of the daye, that is cleped the altytude meridian.

Sette the degre of thy sonne upon the lyne meridionale and reken how many degrees of almicanteras ben betwixe thyn Est orizonte and the degre of thy sonne and take there thin altitude meridian; that is to sain, the highest degre of the sonne as for that daye. So mayest thou knowe in the same lyne the highest degre† that any sterre fyxe clymbeth by nyght: this is to saine that when any sterre fyxe is passed the lyne meridionall, than begynneth it to discende; and so doth the sonne.

14 XVIII. To knowe the degre of the sonne by the rete for a maner curyosyte.

Seke busely with thy rule the highest of the sonne in myddes of the daye; tourne than thyn astrolabie, and with a prycke of ynk marke the nombre of the same altitude in the lyne meridionale. Tourne than thy rete about tyl thou finde a degre of thy zodiake accordyng with the prycke; that is to sain, syttyng on the prycke—and in sothe thou shalt finde but two degrees in al the zodiake of that condycyon, and yet thilke two degrees ben in divers sygnes. Than mayst thou lightly by the seson of the yere knowe the signe in whiche is the sonne.

15 XIX. To knowe which day is like to other in lengthe throughout the yere.

Loke whiche degrees ben lyke [far] from the hedes of Cancer and Capricorn; and loke when the sonne is in any of thilke degrees; than ben the dayes lyke of length; that is to sain—that as long is that day in that moneth as was soche a daye in soche a moneth, there varieth but lyttel. Also if thou take two dayes naturelles in the yere ylike farre from eyther point of the equinoctial in the opposite partyes; than as long is the day artificial on that one day, as on that other; and eke the contrarie.

* Signet, *i.e.*, Zenith; pronounced with the French silent g, and sometimes written *synet* or *sinet* by Chaucer (See page 45).

† In all the copies this word is "*lyne.*" It ought manifestly to be "*degre.*"

16 XX. This chapter is a maner declaracion to conclusions that followeth.

Understande wel that thy zodiake is departed in two* halfe circles, from the hed of Capricorn unto the hed of Cancer, and ayenwarde from the hed of Cancer unto the hed of Capricorn. The hed of Capricorn is the lowest poinct where as the sonne goth in winter, and the hed of Cancer is the highest poinct in which the sonne goth in sommer. And therefore understande wel that any two degrees that ben ylike far from any of these two hedes truste wel that thilke two degrees ben of lyke declinacion, be it southward or northward, and the dayes of hem ben lyke of length, and the nyghtes also; and shadowes ylyke, and the altitudes ylyke at myd-day for ever.

17 XXI. To knowe the verrey degre of any maner sterre strange, after his altitude.† Though he be indeterminate in thyn astrolabye sothely to the trouth thus he shal be knowe.

Take the altitude of thy sterre when he is on the Est syde of the lyne meridional as nygh as thou mayest gesse, and take that ascendant anone right by some maner sterre fyxe which thou know-

---

* *Sic* in MSS. The printed copies have "*into* halfe circles."

† This word is *latitude* in all the copies—an obvious error, since the chief object of the problem is longitude. But as the transposition of the first two letters is the commonest of all errors, and as altitude makes very good sense as one of the principal elements of the problem, I have substituted it as the most probable original. This is one of the Conclusions that Stevins "cleane put out for utterly false and untrew," possibly because he could not understand it, or because it was one of those denounced by Stoeffler. It is true that it is not strictly correct in theory—because a mean of two ascendants is not necessarily the ascendant of the mean of their relative culminating points. But the method is sufficiently correct for practical purposes with an instrument so imperfect in itself as the Astrolabe: provided that the interval between the equal altitudes be short, a condition which Chaucer seems to have had in view when he directs the first altitude to be taken on the east side of the meridian "*as nygh as thou may'st gesse.*" If the interval were not more than an hour the greatest error in the result could not exceed a degree of longitude—much less than might arise from the other sources of error to which such an observation would be exposed. It must be observed that the longitude of which Chaucer speaks was not, as at present, referred to the Pole of the ecliptic; but was that degree of the ecliptic that came to the meridian with the star. This was "*Longitudo secundum cœli mediationem,*" and was still in use' long after Chaucer's time. Also it must be observed that Chaucer makes no distinction between latitude and declination but treats both these terms as synonymous.

est; and forget not the altitude of the firste sterre ne thyn ascendente. And when that thys is done aspye dilygently when this first sterre passeth anythyng to the south westward and cacche him anone ryght in the same nombre of the altitude on the west syde of thys lyne meridional as he was caught on the est syde, and take newe ascendente anone right by some maner sterre fyxe the which that thou knowest, and forgete not this second ascendant. And when this is done, reken thou howe many degrees ben bitwixe the first ascendent and the second ascendent, and reken wel the myddel degre betwix bothe ascendentes; and sette thilke myddel degre upon thyn Est orizonte: and then loke what degre sitte upon the lyne meridional and take there the very degre of the ecliptike in whiche the sterre standeth for the tyme. For in the ecliptike is the longitude of a celestiale body, rekoned even fro the hed of Aries unto the ende of Pisces; and his latitude is rekened after the quantyte of his declinacion north or south toward the poles of this worlde. As thus: if it be of the sonne, or any fyxe sterre, reken his latitude or his declynacion fro the equinoctial circle; and if it be of a planete reken than the quantite of his latitude from the ecliptike lyne. Albeit so that from the equinoctial maye the declinacion or the latitude of any body celestiale be rekened after the sight,* north or south, and after the quantite of his declinacion; and yet so maye the latitude or declynacion of any body celestiall, save only of the sonne, after his sight, and after the quantite of his declinacion, be rekened from the ecliptike lyne, fro which lyne all planetes sometyme decline north or south save only the forsaied sonne.

XXII. To knowe the degrees of longitude of fyxe sterres after that thei be determinate in thin astrolabye if it so be that they ben trewly sette.

---

* It is difficult to interpret "*after the sight.*" It is just possible it may mean according to visual observation: or it may be that sight is a copy-error for *height* = *culmen* = *meridian altitude.* The passage in which it occurs, unless some part of it has been lost, merely declares that latitude or declination may be reckoned from either the *equinoctial* or the *ecliptic*; except in the case of the sun, when it must be from the equinoctial only. Then, reading "declinacion" in that place as simply *deviation*, the sense might be:—The latitude of any celestial body may be reckoned from the equinoctial, north or south, after its sight, and after the quantity of its deviation, &c.

Sette the centre of thy sterre upon the lyne merydional and take kepe of thy Zodiake and loke what degre of any signe sytte upon the same lyne meridional at the same tyme, and take there the degre in which the sterre standeth. And with the same degre cometh the same sterre into the same lyne fro the orizonte.*

22 XXIII. To knowe in special the altitude of our centre, I mene after the latitude of Oxenforde and the hight of our Pole.

Understande wel that as farre is the hed of Aries or Libra in the equinoctial from our orizonte as is the synet from the pole artike; and as hie is the pole artike from the orizonte as the equinoctial is ferre fro the synet. I preve it thus by the latitude of Oxenforde: understande wel that the height of our pole artike from our north orizonte is 51 degrees and 50 minutes; than is the synet from the pole artike 38 degrees and 10 minutes: than is the equinoctial from our sinet 51 degrees and 50 minutes: than is our south orizonte from our equinoctial 38 degrees and 10 minutes.

Understande wel this rekening, also forget not that the sinet is 90 degrees of height from the orizont, and our equinoctial is 90 degrees from our pole artike. Also this shorte rule is sothe, that the latitude of any place† in a region is the distaunce from the sinet unto the equinoctial.

23 XXIV. To prove evidentlye the latytude of any place in a region by the preffe of the hyght of the pole artike in that same place.

In some wynters nyght, when the firmament is clere and thick sterred, wayte a time tyl that any ster fixe sitte line right perpendiculer over the pole artike and clepe that ster A; and wayte another ster that sytte line right under A and under the pole and clepe that ster F: and understande wel that F is not considred but onely to declare that A sytteth over the pole. Take than anone right the altitude of A from the orizonte and forget it

* Stevins, according to his Preface, intended to reject this conclusion also (*supra*, pp. 23, 24); nevertheless it is certainly included in his MS. With the longitude used by Chaucer, as explained in a preceding note, there is nothing in this conclusion to be objected to. The last clause is a necessary consequence of that kind of longitude; since both star and degree lie upon the same circle of declination, which, in the astrolabe, is projected in a straight line from the horizon to the pole.

† *Sic* in Stevins. All other copies have "*planet.*"

not. Let A and F go farewel tyl agaynst the dawning a gret while; and come then agayne and abyde tyll that A is even under the pole and under F; for sothely then wol F sytte over the pole. Take than eftsones the altitude of A from the orizonte and note as wel the second altitude as the first altitude. And when that thys is done reken how manye degrees that the first altitude of A exceeded his second altitude and take halfe the ilke porcion that is exceeded and adde it to his second altitude and take there the elevacion of the pole and eke the latitude of thy region. For these two ben of one nombre, that is to sain, as many degrees as thy pole is elevat so moche is the latitude of thy region. Ensample as thus: Peraventure the altitude of A in the evening is 82 degrees of height, than will the second altitude in the dawning be 21; that is to saine, lesse by 61 than was his first altitude at even. Take than the halfe of 61 and adde to it 21 that was his second altytude, and than hast thou the height of the pole and the latitude of thy region.* But understonde wel, to preve this conclusion, and many another fair conclusion, thou mayst have a plomet hanging on a lyne higher than thy hed, on a perche, and that lyne mote hange even perpendiculer betwixe the pole and thyne eye: and than shalte thou se yf A sytte even over the pole and over F at even, and also if F sytte even over the pole and over A at daye.

XXV. Another conclusion to preve the hyght of the pole artike from the orizont.

---

* The two stars α and β, Ursæ Majoris, still called *the pointers*, were, in Chaucer's time, much more nearly in a direct line with the north pole than at present; and nothing would prevent their being the stars intended by Chaucer, but that they are both at the same side of the pole. One of them, however, might be, and no doubt was, his star A; and F, the star under the pole, was as certainly *Polaris*, at that time nearly four degrees from the pole, and almost exactly in a direct line passing through it to the stars first named. Now the discrepancies in the various copies are confined to the upper altitude: some making it 92 and others 62 degrees; but they all agree in the lower altitude, 21 degrees, and as that is just consistent with β Ursæ Majoris, in the lat. of 51° 30′, it may be safely assumed that the upper altitude ought to be 82 degrees, which I have consequently placed in the text. The respective right-ascensions of these two stars at either side of the pole were 157 and 338 degrees—that is, within less than a degree of being diametrically opposite—and when it is considered that these places have been carried back by tabular corrections through nearly five centuries, and that Chaucer's observation of verticality was by plumb-line and naked eye, the approximation is remarkable and quite sufficient to warrant the correction in the text.

Take any sterre fyxe that never descendeth under the orizont in thilke region, and consyder his hyghest altitude and his lowest altitude from the orizont, and make a nombre of these altitudes; take than and abate halfe that nombre [from his hyghest altitude] and take there the elevacion of the pole artike in that same region.

XXVI. Another conclusion to preve the latitude of a region that ye ben in.

Understande wel that the latytude of any place in a region is verely the space betwyxe the signet of hem that dwellen there and the equinoctiall circle, north, or south, taking the mesure in the merydyonall line as sheweth in the almicanteras of thin astrolaby; and thilke space is as moche as the pole artyke is hie in the same place fro the orizont. And than is the depressyon of the pole antartyke beneth the orizonte the same quantite of space, neither more ne lesse: than, if thou desire to know this latitude of the region, take the altitude of the sonne in the myddle of the daye, when the sonne is in the hed of Aries or of Libra, for than movethe the sonne in the lyne equinoctial, and abate the nombre of that same sonne's altitude out of 90 degrees and than is the remnaunt of the nombre that leveth the altitude of the region: as thus—I suppose that the sonne is thilke daye at noon 38 degrees of heyght, abate, than, 38 degrees out of 90 so leveth ther 52, than is 52 degrees the latitude. I saye not this but for ensample, for wel I wote the latitude of Oxenforde is certain minutes lesse. Nowe if it so be that thu thinketh too long a tarying to abyde til that the sonne be in the hed of Aries or Libra, than waite when that the sonne is in any other degre of the zodiake and consider if the degre of his declinacion be northward from the equinoctial; abate than from the sonne's altytude at none the nombre of his declinacion, and than hast thou the height of the hedes of Aries and Libra: as thus:—my sonne, peraventure, is in the 10 degre of Leo, almost 56 degrees of height at none, and his declinacion is almost 18 degrees northward from the equinoctial; abate than thilke 18 degrees of declinacion out of the altitude at none, than leveth 38 degrees—lo there the height of the hed of Aries or Libra and thyn equinoctial in that region. Also if it so be that the sonne's declinacion be southward from the equinoctial, adde than thilke declinacion to the altitude of the sonne at none and take there the hedes of Aries and Libra and thyn equinoctiall. Abate than the height of the equinoctial out of 90 degrees and than leveth there 52 degrees; that is the distaunce of thè region from the equinoctial. [Or take the highest altitude]* of any sterre fyxe that thou knowest and take [or add his declinacion]* from the same equinoctial lyne, and werke after the maner aforesaid.

XXVII. Declaracion of the ascencions of signes as well in the circle directe as in oblique.

The excellence of the sphere solid, amonges other noble conclusions, sheweth manifest the divers ascencions of signes in divers places, as wel in right circles as in embolyfe circles. [A right circle or horizon have those poeple that dwell under the equinoctial line]† and evermo the arche of the daye and the arche of the night is there ylike longe; and the sonne twise every yere passeth through the signet over hed; and 2 sommers and 2 wynters in a yere have these forsayd peple; and the almycanteras in their astrolabie ben straight as a line. And note that this right orizont—that is cleped orizont *rectum*—devideth the equinoctial into right angles: and the embolife orizont, wheras the pole is enhaunced upon the orizonte, overcometh the equinoctial in embolife angles.

The utilitie to knowe the ascencions of signes in the right circle is this: truste wel that by mediacyons of thilke ascencions, these astrologiens by their table and ther instrumentes knowen verely the ascencion of every degre and minute in al the zodiake in the embolife circle, as shal be shewed. These auctours writen that thilke signe is cleped of right ascencion with which the more part of the circle equinoctiall and the lesse part of the zodiakė ascendeth; and thilke signe ascendeth embolyfe with which the lesse of the equinoctial and the more part of the zodiake ascendeth.

XXVIII. This is the conclusion to knowe the ascensions of signes in the right circle, that is, *circulus directus.*

Sette the hed of what signe thu lyste to knowe the ascendyng on the right circle, upon the lyne meridionall and wayte where thine almurie toucheth the bordure, and set there a prycke: tourne

* The words in the first brackets are added as a necessary connection with the preceding matter; and those in the second brackets as substitutes for these unintelligible words which occupy their place in the MSS. and printed copies—"the nether elongation lengthening."

† I have added these words, in brackets, to supply the evident hiatus: and I have altered the relative places of several of the passages in order to render them more congruous and consecutive.

than thy rete westward til the ende of the forsaide signe sitte upon the meridional lyne; and eftsones wayte where thine almurie toucheth the bordure and sette there another prycke. Reken than the nombre of degrees in the bordure betwixe bothe pryckes, and take than the ascencion of the signe in the right circle; and thus maist thou werk with every porcion of the zodiake.

XXIX. To knowe the ascencions of signes in the embolife circle in every region: I mene *in circulo obliquo.*

Sette the hed of the signe, which as thu liste to knowe his ascencion, upon the est orizonte, and wayte where thine almury toucheth the bordure, and sette there a prycke; tourne than thy rete upwarde til the ende of the same sygne sitte upon the est orizonte and waite eftsones whereas thine almury toucheth the bordure, and sette there another prycke: reken than the nombre of the degrees in the bordure betwixe bothe pryckes and take there the ascension of the signe in the embolyfe circle. And understande wel that alle the signes in the zodiake from the hed of Aries unto the ende of Virgo ben cleped signes of the north from the equinoctial; and these signes arisen betwixe the verray Est and the verray North in oure orizont generally for ever. And alle the signes from the heed of Libra unto the ende of Pisces ben cleped signes of the South fro the equinoctial,—and these signes arisen evermore betwixe the verray Est and the verray South in oure orizont. Also every signe betwixe the hed of Capricorn unto the ende of Gemini ariseth in oure orizonte in lesse than two houres equalles; and these same signes from the hed of Capricorn unto the end of Gemini ben called tortuous signes, or croked signes, for thei arysen embolife in oure orizonte: And these croked signes ben obedient to the signes that ben of the righte ascension. These signes of righte ascension ben fro the heed of Cancer unto the ende* of Sagittary; and these signes arisen more upright than dothe the othere, and therefore they ben called soveraine signes,—and every of hem ariseth in more space than in two houres: of whiche signes Gemini obeyeth to Cancer, and Taurus to Leo, and Aries to Virgo, Pisces to Libra, Aquarius to Scorpio, and Capricorn to Sagittary. And thus evermore two signes that ben like farre from the hed of Capricorn obeyeth every of hem to other.

---

* "Hed" in printed copies.

XXX. To knowe justly the foure quarters of the world, as Est, West, South, and North.

Take the altitude of thy sonne when thou liste, and note well the quarter of the world in which the sonne is from [for] the tyme by the asymutes: tourne then thine astrolaby and sette the degre of the sonne in the almicanteras of his altitude on thilke syde that the sonne standeth, as is in maner of taking of houres; and laye thy labell on the degre of the sonne, and reken howe many degrees of the *bordure** ben bytwene the lyne meridionall and the point of thy label, and note wel the nombres. Tourne than agayne thyne astrolabie and set the point of thy grete rule, there thou takest thin altitudes, upon as many degrees in hys bordure from his meridional as was the pointe of thy labell from the lyne meridional on the wombe syde. Take then thyne astrolabye with bothe handes sadly and slyly and let the sonne shine through bothe holes of thy rule and slyly in thilke shyning laye thine astrolabye couche adoune even upon a playne ground; and than wyl the meridionall lyne of thin astrolabye be even south, and the est lyne will lye even est, and the west lyne west, and the north lyne north, so that thou worke softely and avisely in the couchynge. And thou hast thus the foure quarters of the firmamente.

XXXI. To knowe the altitude† of Planettes from the way of the sonne, whether they ben north or south fro the way aforesaide.

Loke, when a planet is on the lyne meridional, yf that her altitude be of the same height that is the degre of the sonne for that daye and than is the planet in the very way of the sonne and hath no latitude. And yf the altitude of the planette be hyer than the degre of the sonne, than is the planet north from the way of the sonne a quantite of latitude as sheweth by thine almycanteras: and yf the altitude be lesse than the degree of the sonne, than is

---

* Here I have not hesitated to substitute "degrees of the *bordure*," for "degrees of the *sonne*," because it is probably a mistake of the copyist; but I do not attempt to alter the fundamental error, which appears to be Chaucer's own, of taking the sun's hour angle to be equal to his azimuth from the south.

† Stevens reads *latitude*, but the word in the text may stand; for, although the alteration is probably right, yet the whole proposition is so erroneous that

the planette south from the waye of the sonne soche a quantite of latitude as sheweth by thine almicanteras. This is to saine, from the waye of the sonne in everye place of the zodiake, for on the morowe the sonne wyll be in another degre.

XXXII. To knowe the signet* for the arising of the sonne; that is to saine, the partie of the orizonte in which the sonne ariseth.

Thou muste first consider that the sonne ariseth not [alwayes] in the verie Est, but sometyme in the North-Est, and sometyme in the South-Est: sothely the sonne ariseth nevermore in the verie

---

this single correction would be of no importance. It is difficult to believe that Chaucer could have entertained such an extraordinary misconception as is implied in this conclusion; and yet since it is the same in all the copies, and since, moreover, the error seems so consistent throughout that no supposition of misprint or miscopying can account for it, I do not attempt to amend it. Its falseness will be apparent when it is considered that at the summer solstice the meridian altitude of the sun is at its highest, but that at midnight of the same day a planet on the meridian, if in the ecliptic, or way of the sun, so far from being "*of the same heighte that is the degre of the sonne for that daye*," would be 47 degrees lower, or the whole breadth of the tropics. What Chaucer really intended was, no doubt, to compare the altitude of the planet with that of the culminating point of the ecliptic at the time of observation.

It is curious that Stevins should have seen that there was an error, and yet, in an attempt to amend it, should have committed a still greater blunder himself: his note to "way of the sonne" is as follows—"The way of the sonne is taken heere for the circuit that the sonne maketh every day by force of the first movable from east to west; and not for the ecliptike, as in other places it is used." That is, he would have the way of the sun in this proposition to mean a parallel to the equinoctial; which, in the example already mentioned, would be the circle called the tropic of Cancer. But to imagine a planet at midnight in that circle, with the sun in the summer solstice, would be to assign to the planet a latitude from the ecliptic of 47 degrees, a worse blunder than the other. The only way in which it seems possible to render the conclusion correct is by the following addition, in brackets, which, after all, is not very extensive:—"Look, when a planet is in the line meridional, if that her altitude be of the same height that is the sun for that day [*when he is in the same sign and degree that is the planet*], then is the planet in the very way of the sun and hath no latitude."

But, of course, the other parts of the Conclusion should be altered in conformity.

* Wherever *signet* occurs in this or the following conclusions, Stevins alters it to *zenith*; but very improperly (see note to conclusion xvi.). It is worthy of note that *zenith*, in the time of Stevins and Stoeffler, was used with the same double application as the *signet* of Chaucer—not only to the vertex, but also to any point in the horizon or equinoctial.

Est in our orizont but if he be in the hed of Aries or Libra. Now is thine orizont departed into xxiiii parties, or minutes, by thine asymutes; in significacion of xxiiii partes of the worlde: though it be so that shipmen reken all the parties in xxxii.

Then is there no more but waite in the whiche minute that the sonne entreth at his arising and take there the signet of the rising of the sonne.

XXXIII. The maner of devision [of the horizont] of thine astrolabie is thus enjoyned: as in this case.*

First, it is devided in fowre plages principallie, with the line that cometh fro Est to West and then with another line that goeth from South to North: then is it devided in smale parties, or minutes, as Est, and Est by South where that is the first minute above the Este line; and so forth fro partie to partie till that thou come again to the Est line. Thus thou might understande the signet of every sterre—in whiche partie he ariseth.

XXXIV. To knowe in which partie of the firmament is the conjunction.

Consider the tyme of the conjunction, by the Kalender; as thus, how many houres that the conjunction is fro middaye of the daye before, as sheweth the canon of the kalender. Rekene then that nombre in the bordure of thine astrolabie, as thou were wont to doe in knowyng of the houres of the day or of the night, and lay thy labell over the degre of the sonne: then will the poinct

* Stevins makes this conclusion a part of the preceding; to which, indeed, it seems supplementary: but not more so than to the original description of the azimuths at page 28. Chaucer's xxiv azimuths were:—

1. East.	7. South.	13. West.	19. North.
2. E. by S.	8. S. by W.	14. W. by N.	20. N. by E.
3. S.E. by E.	9. S.W. by S.	15. N.W. by W.	21. N.E. by N.
4. S.E.	10. S.W.	16. N.W.	22. N.E.
5. S.E. by S.	11. S.W. by W.	17. N.W. by N.	23. N.E. by E.
6. S. by E.	12. W. by S.	18. N. by W.	24. E. by N.

These azimuths, or minutes, were each of 15 degrees, and their names are indicated by Chaucer himself by calling "East by South—*the first minute above the Est line*—and so forth fro partie to partie till that thou come again to the Est line."

The concluding words in the title of this 33rd conclusion—"as in this case" should probably be, *as in these cases*,—referring to the several propositions respecting azimuths which precede and follow it. In Stevins, *minute* is altered everywhere to *azimuth*, but quite unnecessarily.

of the labell sitte upon the houre of the conjunction. Loke then on which minute the degre of the sonne sitteth, and in that partie of the firmament is the conjunction.

XXXV. To knowe the signet of the altitude of the sonne.

This is no more to saie but any time of the daie take the altitude of the sonne and by the minute in which he standeth thou might se in which partie of the firmament he is: and in the same wise might thou se by night any sterre whether he sitte est, west, or south, or any part betwixe, after the name of the minute in which the sterre standeth.

XXXVI. To know sothlie the longitude of the Mone, or any planette, that hath no latitude, for the tyme, fro the ecliptike line.*

Take the altitude of the Mone and rekene thyne altitude up among thyne almicanteras on which side that the Mone standeth, and sette there a pryck. Take then anone, right uppon the Mone's side, the altitude of any sterre fyxe that thou knowest and sette his centre upon his altitude among thyne almicanteras there the sterre is founden; wayte than of which degre the zodiake is, to which the pryck of the altitude of the Mone [applies]† and take there the degre in which the Mone standeth. This conclusion is verray sothe if the sterres in thine astrolabye standeth after the trouth. Some tretises of the astrolabie maketh none exception whether the Mone have latitude or none, nor whether side of the Mone the altitude of the sterre be found. And note, if the Mone shewe herself by daye than maist thou werche the same conclusion by the sonne as wel as by the sterre fixe.

XXXVII. This is the werching of the conclusion to knowe whether any planet be directe or retrograde.

Take the altitude of any sterre that is cleped a planete and note it wel: anone right take the altitude of some sterre fixe that thou knowest and note it wel also: and come ageyn the thirde or fourthe night next following, for then thou shalte perceve wel the mevyng of the Planet whether he meve forward or bakkward: and waite wel then whan the starre fixe is in the same altitude that

* Slightly altered from the copies, which have "*fro* the tyme *of* the ecliptike line."

† Or, which degree of the zodiac is cut by the almicanter of the Moon's altitude.

she was when thou toke the firste altitude of the foresaid planet and note it wel. For truste wel if so it be that the Planet be in the righte side* of the meridional line so that his second altitude be less than the firste altitude was, thanne is the planet direct: and if he be in the west side in that condicion, thanne is he retrograde. And if so be that this Planet be in the est side when his altitude is take so that the seconde altitude be more than his firste altitude thanne is he retrograde: and if he be in the west side of the line meridionale than is he direct. But the contrary moving of these parties is the cours of the Mone, for, sothelie, the Mone moveth the contrary fro other planettes in her ecliptick line but in none other maner.

[This is another of the conclusions denounced by Stoeffler and omitted by Stevins. The fatal objection to it, apparently overlooked by Chaucer, is that a change of the planet's *declination* might affect its altitude and vitiate the result. In other respects the method is very ingenious. But it is not easy to account for the concluding paragraph, which states that "the Mone moveth the contrary fro other planets in her ecliptick line." The moon's motion in the ecliptic is direct, like that of the sun; and as it is impossible to suppose Chaucer unmindful of that very obvious fact, the passage must be either hopelessly corrupt or else altogether interpolated.]

XXXVIII. The conclusion of equacions of howses after the astrolabie.

Sette the beginning of the degre that ascendeth upon the ende of the viii houre inequale, than will the line of the seconde howse sitte upon the line of Midnight. Remeve then the degre that ascendeth and sette hym upon the ende of the x houre inequale, then will the beginnyng of the thirde howse sitte upon the Midnight line. Bring up again the same degre that ascendeth first, and set hym upon the Est orizonte and then will the beginning of the fowerth howse sitte upon the Midnight line. Take then the nadyr of the degre that ascendeth first and sette hym upon the ende of the ii houre inequale, and then will the beginnyng of the fifte howse sitte upon the Midnight line. Take then the nadire of the ascendent and sette hym upon the ende of the iiii houre inequale and then will the beginnyng of the sixte howse sitte upon the Midnight line. The beginnyng of the seventh howse is nadire of the ascendent; and the beginnyng of the eyghth howse is nadire of the second; and the beginnyng of the nineth howse is nadire of the third; and the beginnyng of the tenth howse is nadire of the

* That is, the *east* side of the meridian. Vide *supra*, page 23.

iiii; and the beginnyng of the 'leventh howse is nadire of the fyfte; and the beginnyng of the xii howse is nadire of the vi howse.

XXXIX. An other maner of equacions of Howses by the astrolabye.

Take thyne ascendente, and then thou haste the foure angles; for wel thou woste that the opposite of thine ascendente, that is to saie the beginnyng of the seventh howse, sitte upon the west orizonte: and the beginnyng of the tenthe howse sitte upon the line meridionall, and his opposite upon the line of midnight. Then laie thy labell upon the degre that ascendeth and rekene then fro the poinct of thy labell all the degrees in the bordure til that thou come to the meridional line, and departe all thilke degrees into thre evene partes and take there the evene porcions of thre other howses. Laie thy labell over every of these thre parties, and then thou might see by the labell in the zodiake the beginning of these thre howses fro the ascendente, that is to saie, the twelveth next above the ascendente, and then the eleventh howse, and the X howse upon the meridional line as I first saied. The same wise werch fro the ascendent doune to the line of midnight; and thus thou haste thre howses, that is to saie, the beginning of the second, the third, and the fowerth howse: than is the nadire of these thre howses the beginnyng of these thre that foloweth [the seventh: and the nadire of the eleventh, and the twelveth, is the beginnyng of the two that followeth the fourth].*

XL. To finde the line meridionale to dwell fixe in any certain place.

Take a round plate of metall, for warpyng the border† the better, and make thereupon a juste compace a little within the bordure; and laie this rounde plate upon an evene ground or some evene stone or on an even stock fixe in the ground, and laie it even

* I have added this passage, in brackets, because after the mention of the 2nd, the 3rd, and the 4th, "the thre howses that foloweth" would be the 5th, 6th, and 7th; but that would be very different from Chaucer's meaning as fully explained in the preceding conclusion.

† Border=broder (broader) *i.e.*, on account of warping, the broader (or thicker) the better. There is an example of broad, applied to thickness of rim, in *broad wheeled wagon.* Stevins also has *broder.*

by a rule. In the centre of the compace sticke an even pinne or a wire upright, the smaller the better, and sette thy pinne or thy wire by a plumme-rule's ende upright even: *and let this pinne be no lenger than three-quarters* of thy diameter of the compace:* and wayte bysyly aboute tenne or eleven of the clocke when the sonne sheweth, when the shadow of the pynne entreth any thing within the circle of the compace one here brede,† and make there a pryck with ynke. Abide then, still waityng on the sonne, after one of the clocke til that the schadowe of the pinne or of the wire passe any thyng out of the circle or compace be it never so little, and sette there a pryck. Take then a compace and mesure even the middle betwixe both pryckes and sette there a pryck. Take then a rule and drawe a strike even fro the pinne unto the middle pryck, and take there the line meridionale for evermore as in the same place: and if thou drawe a crosse [line] over thwarte the compace justlie over the line meridionall, then haste thou Est and West, and per consequens the oppositife, that is, South and North.

XLI. Discripcion of the Meridionale line, and of the longitudes and latitudes of cities and tounes as well as of Climates.

This line meridionale is but a maner discripcion of a line ymagined that passeth upon the Poles of the world, and by the signet over hede: and it is cleped the signet, for in what place that any man is at any time of the yere whan the sonne by mevyng of the firmament cometh to his meridianale place, thanne is it the verray middaie that we clepe None, and therefore is it cleped the line of Middaie. Than take hede that evermore of two citees, or of two tounes, of whiche the one approcheth neerer the Est than doth the other toune, truste wel that thilke two tounes have diveres meridians. Take kepe also that the arche of the Equinoctial that is conteyned and bounded betwene the

* I have here substituted *three-quarters* for "a quarter," and omitted "fro the pinne," in the words "thy diameter of the compace *fro the pinne,*" which would mean the radius. Now a quarter of the radius would be an eighth of the diameter, a preposterous proportion. But with 9 to 6 as the ratio of the gnomon to the shadow, as I have made it, the altitude of the sun is 56° 18′ about the summer solstice at an hour and a half, or an hour and three-quarters, before noon; which very well agrees with Chaucer's ten or eleven of the clock, and was, beyond all reasonable doubt, what he intended. [Stevins omits altogether the words in italics, and so avoids the difficulty.]

† One hair's breadth.

two meridians is cleped the longitude of the Toune. And if it so be that two tounes have meridian y-like, or one meridian, thanne is the distaunce of hem bothe like ferre.* And in this maner thei chaunge not the meridian, but sothlie thei chaunge ther alimancanteras, for the haunsing of the Pole and the distaunce of the sonne.

The longitude of a clymat maye be cleped the space of the yerth fro the beginnyng of the firste clymat unto the laste ende of the same climat even direct against the pole artike; thus saie some aucthours.†

And some clerkes saie that if men clepe the latitude of a contre‡ the arche meridian that is conteined or intercepte betwixe the signet and the Equinoccial, then, thei saie, that the distaunce fro the Equinoccial unto the ende of the clymate even ayenst the pole artike is the longitude of the clymate fro south.

[Here I think the astronomical conclusions to be attributed to Chaucer himself ought properly to end. As to the one other that remains, the XLIInd, I am convinced it was not written by him, and I have fully explained my reasons for that opinion in the Introduction (p. 11). I there fore make no attempt to amend it. But I shall print it verbatim from Urry's edition: giving it, however, the benefit, such as it is, of Stevins' alterations and additions by noting them at foot of each page.]

(XLII.) To knowe with what degree of the Zodiake that any Planet ascendeth on the orizonte where his latitude be North or South.

Knowe by thyne Almanacke the degre of the Ecliptike of any signe in which that the planette is rekened for to be, and that is cleped the degre of his longitude. And knowe also the degre of his latitude fro th' ecliptike North or South, and by these ensamples folowing in especialle thou maieste wirche with every signe of the Zodiake. The longitude peraventure of Venus, or of an other

* An important part of the text appears to be wanting here, which if not recovered from other MSS., will be much to be regretted—as depriving us of a knowledge of what Chaucer considered to be the first meridian—whether the Gaditanæ or the Fortunate Isles.

† Stevins very improperly changed "longitude" into *latitude* at the beginning of this sentence, for in the next paragraph Chaucer fully explains in what sense *longitude* is to be understood.

‡ This word is *center* in the printed copies.

planet was[1] of Capricorne, and the latitude of 'hem Northward [2] degrees fro the Ecliptike line, then toke I subtil compas and cleped the one poinct of my compace A, and that other F, then toke I the poinct of A and set in the ecliptike line[3] and my Zodiake in the degre of the Longitude of heddes[4] that is to saie in the ende[5] of Capricorne, and then set I the poinct of F upward in the same signe bicause that the altitude[6] was North upon the latitude of Venus, that is to saie in the[7] degre fro the hed of Capricorne, and thus have I the[8] degrees betwixe my two prickes[9], then laied I downe softlie my compace and set the degre of the longitude upon the[10] Orizont, then toke I and waxed my labell in maner of a paire of tables to receve distinctly the pricke of my compace, then toke I this foresaid labell and laid it fixe over the degre of my longitude, then toke I up my compace and[11] the poinct of A in the waxe of my labell as I coud gesse over th' ecliptike line in th' ende[12] of the longitude. I set the poinct[13] over endlonge on the labell upon the space of the latitude inwarde, and on the Zodiake, that is to say Northward fro the ecliptike: (then laide I doune my compace and loked well in the waie upon th' ecliptike of A and F)[14] then tourned I my rete till that the pricke of F sate upon the orizont, then sawe I well that the bodye of Venus in her[15] latitude of degrees septentrionale ascendeth (in the ende of degre)[16] fro the hed of Capricorne. And note that in this maner thou mightest werch with any latitude septentrionall in al signes: but sothly the latitude Meridionall of a planet in Capricorn maie not be take bicause of the little space betwixe the ecliptike and the bordure of the Astrolabie, and sikerly in al other signes it maie be take. Also the degre peraventure of Jupiter, or of any other planette was in the first degre of Pisces in longitude and his latitude was[17] degrees Meridionall. Then toke I the poinct of A and set it in the first degree of Pisces in th' ecliptike, then set I the poinct downward of F in the same signe[18] bicause that the latitude was south[19]

---

[1] Here Stevins inserts "*in the hedd.*" [2] Inserts "2." [3] for "*and,*" substitutes "*of.*" [4] demit "*of heddes.*" [5] for "*ende,*" substitutes "*hedd.*" [6] corrects by erasure to "*latitude.*" [7] in serts "*Seconde.*" [8] for "*the*" substitutes "*two.*" [9] substitutes "*pointes.*" [10] inserts "*east.*" [11] inserts "*sette.*" [12] for "*ende*" substitutes "*degree.*" [13] inserts "*of F,*" and demit "*over.*" [14] demit all within parenthesis. [15] for "*her*" substitutes "*this.*" [16] demit all within parenthesis, and substitutes "*with the ende of Sagittarre a certaine of degrees.*" [17] inserts "3." [18] for "*same*" substitutes "*nexte.*" [19] inserts

degrees, that is to saie fro the hed of Pisces, and thus have I [20] degrees betwixe both prickes. [21] Then set I the degre of the longitude upon the orizont, then toke I my labell and laied him fixe upon the degre of longitude, then set I the poinct of A on my labell even over the ecliptike line on the ende of the [22] degre of the longitude and I sette the poinct of F endelong on my labell the space of [23] degrees of the latitude outwarde fro the Zodiake, that is to saie southwarde fro the Ecliptike towarde the bordure, and then tourned I my rete, til the poincte of F sate upon the orizont, then sawe I well that the bodie of Jupiter in his latitude of [24] degrees meridionall, ascendeth with the [25] degre of Pisces in horoscopo, And in this maner thou maiest werche with any Latitude, as I saied first, save in Capricorne.

And thou wilte plie this crafte with the arysing of the Mone, loke thou reken well the [26] course of [27] houre by houre, for she dwelleth in a degre of her Longitude but a little while, as thou woste well: but neverthelesse if thou legen well her verie mevyng by the tables, or after her course, houre by houre, thou shalte doe well inough.

[The Pratike of Umbra Recta and Umbra Versa: To knowe the heyght of toures and other thynges by the Scale on the bakk-syde of thyne astrolabye.]*

1. Umbra Recta. If thou might come to the base of the towre in this maner shalt thou werke. Take the altitude of the towre with both holes, so that the rule lie even on a poinct. Ensample as thus: thou seest hym through the poinct of fower: then mete I the space betwixe thee and the towre, and I find it twentie fote: then beholde I how fower is to twelve and I find it is the thirde parte of twelve: right so, the space betwixe thee and the towre is

"3." [20] inserts "3." [21] substitutes "*pointes.*" [22] demit "*ende of the.*" [23] inserts "3." [24] inserts "3." [25] inserts "6th." [26] for "*the*" substitutes "*her.*" [27] demit "*of.*"

* I have no authority for these few lines of introductory title, but they seem absolutely necessary to break the abruptness of the change from the astronomical portion; and there can be no doubt that some such introductory matter must have formed part of the original.

the thirde part of the altitude of the towre: then thrise twentie fote is the highest of the towre, with the addicion of thine owne bodie fro thyne eye. If the rule fall on five then [as five is to twelve so is the space betwixe to]* the heyght of the towre.

42 II. UMBRA VERSA. If thou maiest not come to the base of the towre and thou fixe hym through the nombre of one; sette there a pryck at thy fote: then go nere the towre and se hym through the poinct of two, and sette there an other pryck: and then beholde howe one hath hym to twelve, and thou shalte finde that he hath hym twelve sithes: then behold how two have hym to twelve and thou shalte finde it sixe sithes, [the difference is sixe] and therefore the space betwixe the two pryckes is sixe times thyne altitude. And note that at the first altitude of one, thou settest a pryck; and afterward when thou seest hym through at two, there thou settest a pryck; then thou findest betwene lx fote: then thou shalte finde that tenne is the sixth part of sixty: then is tenne fote the altitude of the towre. But if it fall upon an other poinct, as thus: it falleth on sixe at the seconde takyng, when [at the first] it falleth on three: then shalt thou finde that sixe is the second part of twelve, and three is the fowrthe parte of twelve: [the difference is two] that is to saie the space betwixe the two pryckes is twise the heyght of the towre. And if the difference were three, then would it be three times the heythe. *Et sic de singulis.*†

43 III. UMBRA RECTA. An other maner of werching by umbra recta. If thou maiest not come by the base of the towre, wirche in this wise: sette thy rule upon one, tyll thou se the altitude, and set at thy fote a prycke; and then set thy rule upon two, and so doe in the same maner: then loke what is the difference betwixe one and two, and thou shalte finde that it is one: then mesure the space betwixe the two pryckes, and that is the twelveth part of the altitude of the towre—and so of all other.

IV. UMBRA RECTA. If thy rule fall upon the eighte poinct, on

* These words in brackets I have substituted for "*is five times twelve*" of the printed copies, which are obviously erroneous. Stevins gets over the difficulty by omitting the last part altogether.

† The numbers and measurements in the foregoing problem are sadly mangled in the various copies. I have inserted such numbers and made such alterations (denoted by brackets) as seem nearest to the requirement of the context and yet are consistent with correctness.

the right shadowe, then make the figure of 8: then loke howe moche space of the fete is betwixe thee and the towre, and multiplie that by twelve: and when thou hast multiplied it by that same nombre, than devide it by the nombre of eighte and kepe the residue, and adde thy heyght unto thyne eye to the residue and that shall be the verie heighte of the towre. And thus maiest thou worche on the same side from one to twelve.

V. Umbra Recta. Another maner of werkyng upon the same side. Loke upon what poinct thy rule falleth when thou seest the toppe of the towre through the two holes, and then mete the space from thy fote to the base of the towre: and right as the nombre of the poinct hath himself to twelve, right so the mesure betwixt thee and the towre hath hymself to the height of the same towre. Ensample as thus: I sette the case thy rule fall upon eight, then is eight two thirde partes of twelve, so is the space two thirde partes of the towre.

VI. Umbra Versa. To knowe the heyght by the poinct of Umbra Versa. If the rule fall upon iii when thou seest the toppe of the towre, sette a pryck there thy fote standeth, and go nere tyl thou maiest se the same toppe at the poinct of iiii, and sette there an other pryck; then mete how many fote is betwixe the two pryckes, and the height up to thine eye, and that shall be the the height of the towre. And note that iii is the fowerth part of xii, and iiii is the thirde parte of xii. Nowe passeth iiii the nombre of iii by distaunce of i, therefore the same space, with thy height to thy eye, is the height of the towre. And if it were so that there were two or three distaunces in the nombres, so should the mesure betwixe the pryckes be twise or thrise the height of the towre.

VII. Umbra Recta. To knowe the height if thou maiest not come to the base of the thynge. Set thy rule upon what poinct thou wilte, so that thou maist se the toppe of the thinge through the two holes, and make a mark there thy fote standeth, and go nere or ferther tyl thou maieste se it through an other poinct, and make there an other marke: and loke what difference is betwixe the two poinctes in the scale: and right as that difference hath hym to xii, right so the spaces betwixe the two markes hath hym to the height of the thing. Ensample: I sette the case that thou seest it through the poinct of iiii, and after at the poinct of iii. Nowe passeth the nombre of iiii the nombre of iii the distance of i: and right as this difference of one hath hymself to xii, right so the mesure betwixe bothe the markes hath hym to the height of the same thing, puttyng thereto the height of thyself to thine eye. And thus maiest thou werke from i to xii.

VIII. Umbra Versa. Furthermore, if thou wilte knowe in umbra versa by the crafte of umbra recta, I suppose thou takest thine altitude at the poinct of iiii and makest a mark, and then thou goest nere* tyl thou haste it at the poincte of iii and makest there an other mark; then must thou devide 144 by 4; the nombre that cometh thereof shall be 36; and after, divide 144 by 3, and the nombre that cometh thereof is 48: then loke what difference is betwixte 36 and 48, and that shalte thou finde is 12: and right as 12 hath hym to 12 so the space betwixte the two pryckes hath hym to the altitude of the thynge.†

---

* Stevins, not understanding this problem, changed "*nere*" into "*further.*"

† In this last problem, what Chaucer means by "the crafte" of umbra recta, is what we should call the stating, or the order in which the terms of the proportion are presented. In umbra recta the first term is the difference between the observed points of the scale, the second is the distance between the stations, the third is the length of the scale, and the fourth is the altitude of the object required; but in Umbra Versa the two first of these terms are transposed. Therefore, in order to take the altitude by the points of umbra versa, and yet perform the computation by the proportion of umbra recta, it is necessary to convert the points of the one into equivalent points of the other: and this is done reciprocally in both scales by dividing the points of each into the square of its whole scale: or, as Chaucer has it, into 144, the square of 12.

Ende of Seconde Partye.

(*Cæteræ Desunt.*)

Appendix.

REPRINTS OF PAPERS ON THE

ASTRONOMY OF CHAUCER

IN "THE CANTERBURY TALES."

WITH ADDITIONAL NOTES.

Essays

ON THE MEANING OF CHAUCER'S PRIME,

ON THE CARRENARE,

ON SHIPPES OPPOSTERES.

I.

THE PILGRIMAGE TO CANTERBURY.

"Whan that Aprille with his shoures soote
The drought of March hath perced to the roote
And bathed every veyne in such licour
Of which vertu engendred is the flour:—
When Zephirus eek with his swete breeth
Inspired hath in every holt and heeth
The tendre croppes, and the yonge Sonne
Hath in the Ram his halfe cours y-ronne:
* * * * *
Thanne longen folk to gon on pilgrimages
* * * * *
* * * * *
Bifel that in that seson on a day."

PROLOGUE.

I quote these lines because I wish to show that Tyrwhitt, in taking them as indicative of the very day on which the journey to Canterbury was performed, committed a great mistake.

The whole of the opening of the prologue, down to the line last quoted, is descriptive, not of any particular day, but of the usual season of pilgrimages; and Chaucer himself plainly declares, by the words "in *that* season, on *a* day"—that the day is *as yet* indefinite.

But because Tyrwhitt, who, although an excellent literary critic, was by no means an acute reader of his author's meaning, was incapable of appreciating the admirable combination of physical facts by which Chaucer has not only identified the real day of the pilgrimage, but has placed it, as it were, beyond the danger of alteration by any possible corruption in the text, he set aside these physical facts altogether, and took in lieu of them the seventh and eighth lines of the prologue quoted above, which I contend, Chaucer did not intend to bear any reference to the day of the journey itself, but only to the general season in which it was undertaken.

But Tyrwhitt, having seized upon a favourite idea, seems to have been determined to carry it through at any cost, even at that of altering the text from "*the Ram*" into "*the Bull*:" and I fear that he can scarcely be acquitted of unfair and intentional misquotation of Chaucer's words, by transposing "his halfe cours" into "half his course." which is by no means an equivalent expression. Here are his own words:

"When he (Chaucer) tells us that 'the shoures of April had *perced to the rote* the drought of March' (ver. 1, 2), we must suppose, in order to allow due time for such an operation, that April was far advanced; while, on the other hand, the place of the sun, 'having just run *half his course in the Ram*' (ver. 7, 8), restrains us to some day in the very latter end of March. This difficulty may, and, I think, should, be removed by reading in ver. 8, the BULL, instead of the RAM. All the parts of the description will then be consistent with themselves, and with another passage (ver. 4425), where, in the best MSS., *the eighte and twenty* day of April is named as the day of the journey to Canterbury."—*Introductory Discourse.*

Accordingly, Mr. Tyrwhitt did not hesitate to adopt in his text the twenty-eighth of April as the true date, without stopping to examine whether that day would, or would not, be inconsistent with the subsequent phenomena related by Chaucer.

Notwithstanding Tyrwhitt's assertion of a difficulty only removable by changing the Ram into the Bull, there are no less than two ways of understanding the seventh and eighth lines of the prologue so as to be perfectly in accordance with the rest of the description. One of these would be to suppose the sign Aries divided into two portions (not necessarily *equal* in the phraseology of the time), one of which would appertain to March, and the other to April—and that Chaucer, by "in the Ram his halfe cours," meant the *last*, or *the April* half of the sign Aries. But I think a more probable supposition still would be to imagine the month of April, of which Chaucer was speaking, to be divided into two "halfe cours," in one of which the sun would be in Aries, and in the other in Taurus; and that when Chaucer says that "the yonge Sonne had in the Ram his halfe cours yronne," he meant that the *Aries half of the month of April* had been run through, thereby indicating *in general terms* some time approaching to the middle of April.

Both methods of explaining the phrase lead eventually to the same result, which is also identical with the interpretation of Chaucer's own contemporaries, as appears in its imitation by Lydgate in the opening of his "Story of Thebes:"—

"Whan bright Phebus passed was the Ram,
Midde of Aprill, and into the Bull came."

And it is by no means the least remarkable instance of want of perception in Tyrwhitt, that he actually cites these two lines of Lydgate's *as corroborative of his own interpretation,* which places the sun *in the middle of Taurus.*

I enter into this explanation, not that I think it necessary to examine too curiously into the consistency of an expression which evidently was intended only in a general sense, but that the groundlessness of Tyrwhitt's alleged necessity for the alteration of "the Ram" into "the Bull" might more clearly appear.*

I have said that Tyrwhitt was not a competent critic of Chaucer's practical science, and I may perhaps be expected to point out some other instance of his failure in that respect than is afforded by the subject itself. This I may do by reference to a passage in "The Marchante's Tale," which evinces a remarkable want of perception not only in Tyrwhitt, but in all the editors of Chaucer that I have had an opportunity of consulting.

The morning of the garden scene is said in the text to be "er that days eight were passed of the month of *Juil*"—but a little further on, the same day is thus described :—

> "Bright was the day and blew the firmament,
> Phebus of gold his stremes doun hath sent
> To gladden every flour with his warmnesse ;
> He was that time in Geminis, I gesse,
> But litel fro his declination
> In Cancer."

How is it possible that any person could read these lines and not be struck at once with the fact that they refer to the 8th of *June* and not to the 8th of *July?* The sun would leave Gemini and and enter Cancer on the 12th of June; Chaucer was describing the 8th, and with his usual accuracy he places the sun "but litel fro" *the summer solstice !*

Since "Juil" is an error common perhaps to all previous editions, Tyrwhitt might have been excused for repeating it, if he had been satisfied with only that: but he must signalise *his edition* by inserting in the Glossary attached to it—"JUIL, *the month of July,*" referring, as the sole authority for the word, to this very line in question of "The Marchante's Tale!"

Nor does the proof, against him in particular, end even there; he further shows that his attention must have been especially drawn to this garden scene by his assertion that Pluto and Proser-

* See Note A at the end.

pine were the prototypes of Oberon and Titania; and yet he failed to notice a circumstance that would have added some degree of plausibility to the comparison, namely, that Chaucer's, as well as Shakspeare's, was a *Midsummer Dream.*

It is, perhaps, only justice to Urry to state that *he* appears to have been aware of the error that would arise from attributing the sun's presence in the sign Gemini to the month of July. The manner in which the lines are printed in *his* edition is this:—

> "ere the dayis eight
> Were passid, er' the month July befill."

It is just possible to twist the meaning of this into *the eighth of the Kalends of July*, by which the blunder would be in some degree lessened; but, inasmuch as the sun, in Chaucer's time, would leave Gemini on the 12th of June, and the eighth of the Kalends of July was not till the 24th of June, such a reading would be as foreign to Chaucer's astronomy as the lines themselves are to his poetry.

[Pub. April 26, 1851.]

II.

THE ARKE OF ARTIFICIAL DAY.

Before proceeding to point out the indelible marks by which Chaucer has, as it were, stereotyped the true date of the journey to Canterbury, I shall clear away another stumbling-block, still more insurmountable to Tyrwhitt than his first difficulty of the "halfe cours" in Aries, viz., the seeming inconsistency in statements (1.) and (2.) in the following lines of the prologue to the Man of Lawe's tale:—

> (1.) "Oure hoste saw wel that the bright sonne,
> The arke of his artificial day, had ironne
> The fourthe part and halfe an houre and more,
> * * * * *
> (2.) And saw wel that the shadow of every tree
> Was as in length of the same quantitie,
> That was the body erecte that caused it,
> And therefore by the shadow he toke his wit
> That Phebus, which that shone so clere and bright,
> Degrees was five and fourty clombe on hight,
> And for that day, as in that latitude
> It was ten of the clok, he gan conclude."

The difficulty will be best explained in Tyrwhitt's own words :—

"Unfortunately, however, this description, though seemingly intended to be so accurate, will neither enable us to conclude with the MSS. that it was '*ten of the clock*,' nor to fix upon any other hour; as the two circumstances just mentioned are not found to coincide in any part of the 28th, or of any other day of April, in this climate."—*Introductory Discourse*, § xiv.

In a foot-note, Tyrwhitt further enters into a calcnlation to show that, on the 28th of April, the fourth part of the day and half an hour and more (even with the liberal allowance of a quarter of an hour to the indefinite phrase "*and more*") would have been completed by nine o'clock A.M. at the latest, and therefore at least an hour too soon for coincidence with (2.).

Now, one would think that Tyrwhitt, when he found his author relating facts, "*seemingly intended to be so accurate*," would have endeavoured to discover whether there might not be some hidden meaning in them, the explaining of which might make that consistent, which, at first sight, was apparently the reverse.

Had he investigated with such a spirit, he must have discovered that the expression "arke of the artificial day" *could not*, in this instance, receive its obvious and usual meaning of the horary duration from sunrise to sunset—

And for this simple reason: That such a meaning would *presuppose a knowledge of the hour*—of the very thing in request—and which was about to be discovered by "our hoste," who "toke his wit" from the sun's altitude for the purpose! But if he knew already that the fourth part of the day IN TIME had elapsed, he must necessarily have also known what that time was, without the necessity of calculating it!

Now Chaucer, whose choice of expression on scientific subjects is often singularly exact, says, "Our hoste *saw* that the sonne," &c.; he must therefore have been referring to some visible situation: because, afterwards, when the time of day has been obtained from calculation, the phrase changes to "*gan conclude*" that it was ten of the clock.

It seems, therefore, certain that, even setting aside the question of consistency between (1.) and (2.), we must, *upon other grounds*, assume that Chaucer had some meaning in the expression "arke of the artificial day," different from what must be admitted to be its obvious and received signifination.

To what other ark, then, could he have been alluding, if not to the *horary* diurnal ark?

I think, to the AZIMUTHAL ARCH OF THE HORIZON included between the point of sunrise and that of sunset!

The situation of any point in that arch is called its bearing; it is estimated by reference to the points of the compass; it is therefore *visually* ascertainable: and it requires no previous knowledge *of the hour* in order to determine when the sun has completed the fourth, or any other, portion of it.

Here, then, is *primâ facie* probability established in favour of this interpretation. And if upon examination, we find that it also clears away the discrepancy between (1.) and (2.), probability becomes certainty.

Assuming upon evidence which I shall hereafter explain, that the sun's declination, on the day of the journey, was 13° 26′ North, or thirteen degrees and a half,—the sun's bearing at rising, in the neighbourhood of London, would be E.N.E., at setting W.N.W.; the whole included arch, 224°; and the time at which the sun would complete one-fourth, or have the bearing S.E. by E., would be about 20 minutes past nine A.M.,—thus leaving 40 minutes to represent Chaucer's "half an hour and more!"

A very remarkable approximation—which converts a statement apparently contradictory, into a strong confirmation of the deduction to be obtained from the other physical facts grouped together by Chaucer with such extraordinary skill!

On the other hand, it is impossible to deny that the "hoste's" subsequent admonition to the pilgrims to make the best use of their time, warning them that "the fourthe partie of this day is gon," seems again to favour the idea that it is the day's actual horary duration that is alluded to.

[Pub May 3, 1851.]

POSTCRIPT IN 1869.

The necessary study of Chaucer's Treatise on the Astrolabe, which I have gone through in editing it, has enabled me to suggest an explanation of this seeming anomaly. In observing "the arche of the day that some folk callen the day artificiale."—Conclusion VIII, page 31. Chaucer would see at the same time the sun's azimuth at rising and setting. And he would take the fourth of the included arch in azimuth as the fourth of the day in time, because it appears by Conclusion XXX, page 51, that he was under the mistaken idea that the azimuth is equal to the hour-angle.

III.

ASTRONOMICAL EVIDENCE.

Unless Chaucer had intended to mark with particular exactness the day of the journey to Canterbury, he would not have taken such unusual precautions to protect his text from ignorant or careless transcribers. We find him not only recording the altitudes of the sun, at different hours, in words; but also corroborating those words by associating them with physical facts incapable of being perverted or misunderstood.

Had Chaucer done this in one instance only, we might imagine that it was but another of those occasions, so frequently seized upon by him, for the display of a little scientific knowledge; but when he repeats the very same precautionary expedient again, in the afternoon of the same day, we begin to perceive that he must have had some fixed purpose; because, as I shall presently show, it is the repetition alone that renders the record imperishable.

But whether Chaucer really devised this method for the express purpose of preserving his text, or not, it has at least had that effect —for while there are scarcely two MSS. extant which agree in the verbal record of the day and hours, the physical circumstances remain, and afford at all times independent data for the recovery or correction of the true reading.

The day of the month may be deduced from the declination of the sun; and, to obtain the latter, all the data required are,

1. The latitude of the place.
2. Two altitudes of the sun at different sides of noon.

It is not absolutely necessary to have any previous knowledge of the hours at which these altitudes were respectively obtained, because these may be discovered by the trial method of seeking two such hours as shall most nearly agree in requiring a declination common to both at the known altitudes. Of course it will greatly simplify the process if we furthermore know that the observations must have been obtained at some determinate intervals of time, such, for example, as complete hours.

Now, in the Prologue to the "Canterbury Tales," we know that that the observations could not have been recorded accept at complete hours, because the construction of the metre will not admit the supposition of any parts of hours having been expressed.

We are also satisfied that there can be no mistake in the altitudes, because nothing can alter the facts, that an equality between the length of the shadow and the height of the substance can only subsist at an altitude of 45 degrees; or that an altitude of 29 degrees (more or less) is the nearest that will give the ratio of 11 to 6 between the shadow and its gnomon.

With these data we proceed to the following comparison:

Forenoon altitude 45°.		*Afternoon altitude* 29°	
Hour.	Declin.	Hour.	Declin.
XI A.M.	8° 9′N.	II P.M.	3° 57′ S.
X „	13° 27′ „	III „	3° 16′ N.
IX „	22° 34′ „	IV „	13° 26′ „
VIII „	Impossible.	V „	Impossible.

Here we immediately select "X A.M." and "IV P.M." as the only two items at all approaching to similarity; while, in these, the approach is so near that they differ by only a single minute of a degree!

More conclusive evidence therefore could scarecely exist that these were the hours intended to be recorded by Chaucer, and that the sun's declination, designed by him, was somewhere about thirteen degrees and a half North.

Strictly speaking, this declination would more properly apply to the 17th of April, in Chaucer's time, than to the 18th; but since he does not profess to critical exactness, and since it is always better to adhere to written authority, when it is not grossly and obviously corrupt, such MSS. as name the 18th of April ought to be respected; but Tyrwhitt's "28th," which he states not only as the result of his own conjecture but as authorised by "the best MSS.," ought to be scouted at once.

In the latest edition of the "Canterbury Tales" (a literal reprint from one of the Harl. MSS., for the Percy Society, under the supervision of Mr. Wright), the opening of the Prologue to "The Man of Lawes Tale" does not materially differ from Tyrwhitt's text, excepting in properly assigning the day of the journey to "the eightetene day of April;" and the confirmation of the forenoon altitude is as follows:—

> "And sawe wel that the schade of every tree
> Was in the lengthe the same quantite,
> That was the body erecte that caused it."

But the afternoon observation is thus related :—

> "By that the Manciple had his tale endid,
> The sonne fro the southe line is descendid
> So lowe that it nas nought to my sight,
> Degrees nyne and twenty as in hight.
> *Ten* on the clokke, it was as I gesse,
> For eleven foote, or litil more or lesse,
> My schadow was at thilk time of the yere,
> Of which feet as my lengthe parted were,
> In sixe feet equal of proporcioun."

In a note to the line "Ten on the clokke" Mr. Wright observes,

"*Ten.* I have not ventured to change the reading of the Harl. MS., which is partly supported by that of the Lands. MS., *than.*"

If the sole object were to present an exact counterpart of the MS., of course even its errors were to be respected: but upon no other grounds can I understand why a reading should be preserved by which broad sunshine is attributed to ten o'clock at night! Nor can I believe that the copyist of the MS., with whom the error must have originated, would have set down anything so glaringly absurd, unless he had in his own mind some means of reconciling it with probability. It may, I believe, be explained in the circumstance that "ten" and "four," in horary reckoning, were *convertible terms.* The old Roman method of naming the hours, wherein noon was the sixth, was long preserved, especially in conventual establishments: and I have no doubt that the English idiomatic phrase "o'clock" originated in the necessity for some distinguishing mark between hours of the clock reckoned from midnight, and hours of the day reckoned from sunrise, or more frequently from six A.M. With such an understanding, it is clear that *ten* might be called *four*, and *four ten*, and yet the same identical hour be referred to; nor is it in the least difficult to imagine some monkish transcriber, ignorant perhaps of the meaning of "o'clock," might fancy he was correcting, rather than corrupting, Chaucer's text, by changing "foure" into "ten."

I have, I trust, now shown that all these circumstances related by Chaucer, so far from being hopelessly incongruous, are, on the contrary, harmoniously consistent;—that they all tend to prove that the day of the journey to Canterbury could not have been later than the 18th of April;—that the times of observation were certainly 10 A.M. and 4 P.M.;—that the "arke of his artificial day" is to be understood as the horizontal or azimuthal arch;—and that the "halfe cours in the Ram" alludes to the completion of the last twelve degrees of that sign, about the end of the second week in April.

There yet remains to be examined the signification of those three very obscure lines which immediately follow the description, already quoted, of the afternoon observation:

> "Therewith the Mones exaltacioun
> In mena Libra, alway gan ascende
> As we were entryng at a townes end."

It is the more unfortunate that we should not be certain what it was that Chaucer really did write, inasmuch as he probably intended to present, in these lines, some means of identifying the year, similar to those he had previously given with respect to the day.

When Tyrwhitt, therefore, remarks, "In what year this happened Chaucer does not inform us"—he was not astronomer enough to know that if Chaucer had meant to leave, in these lines, a record of the moon's place on the day of the journey, he could not have chosen a more certain method of informing us in what year it occurred.

[*Pub. May* 17, 1851.]

IV.

THE STAR MIN AL AUWÂ.

> "Adam Scrivener, if ever it thee befall
> Boece, or Troilus, for to write new,
> Under thy long locks mayst thou have the scall
> But, after my making, thou write more trew;
> So oft a day I mote thy worke renew,
> It to correct, and eke to rubbe and scrape,
> And all thorow thy negligence and rape."
>
> *Chaucer to his own Scrivener.*

If, during his own lifetime, and under his own eye, poor Chaucer was so sinned against as to provoke this humorous malediction upon the head of the delinquent, it cannot be a matter of surprise that, in the various hands his text has since passed through, many expressions should have been perverted, and certain passages wholly misunderstood. And when we find men, of excellent judgment in other respects, proposing, as Tyrwhitt did, to alter Chaucer's words to suit their own imperfect comprehension of his meaning, it is only reasonable to suspect that similar mistakes may have induced early transcribers to alter the text wherever, to their wisdom, it may have seemed expedient.

Now I know of no passage more likely to have been tampered with in this way than those lines of the prologue to the *Persone's Tale,* alluded to at the close of my last communication. Because, supposing (which I shall afterwards endeavour to prove) that Chaucer really meant to write something to this effect: "Thereupon, as we were entering a town, the moon's rising, with Min al auwâ in Libra, began to ascend (or to become visible),"—and supposing that his mode of expressing this had been,

> "Therewith the mone's exaltacioun,
> In libra men alawai gan ascende,
> As we were entrying at a towne's end :"

—in such a case, what can be more probable than that some ignorant transcriber, never perhaps dreaming of such a thing as the Arabic name of a star, would endeavour *to make sense* of these, to him, obscure words, by converting them into English. The process of transition would be easy; "min" or "men" requires little violence to become "mene" (the modern "mean" with its many significations), and "al auwâ" (or "alwai," as Chaucer would probably write it) is equally identical with "alway." The misplacement of "Libra" might then follow as a seeming necessity; and thus the line would assume its present form, leaving the reader to understand it, either with Urry, as, "I mene Libra," that is, I *refer to* Libra;—or with Tyrwhitt: "In mene Libra," that is, in *the middle of* Libra.

Now, to Urry's reading, it may be objected that it makes *the thing ascending* to be Libra, and does not of necessity imply the moon's appearance above the horizon. But since the rising of the moon is a *visible* phenomenon, while that of Libra is theoretical, it must have been *to the former* Chaucer was alluding, as to something witnessed by the whole party as they

> "Were entrying at a towne's end;"

or otherwise this latter observation would have no meaning.

The objection to Tyrwhitt's reading is of a more technical nature —the moon, if in *the middle* of Libra, *could not* be above the horizon, in the neighbourhood of Canterbury, at four o'clock P.M., in the month of April. Tyrwhitt, it is true, would probably smooth away the difficulty by charging it as another inconsistency against his author; but I—and I hope by this time such readers of these notes as are interested in the subject—have seen too many proofs of Chaucer's competency in matters of science, and of his commentator's incompetency, to feel disposed to concede to the latter such a convenient method of interpretation.

But there is a third objection common to both readings—that they do not satisfactorily account for the word "alway;" for although Tyrwhitt endeavours to explain it by *continually*, "was *continually* ascending," such a phrase is by no means intelligible when applied to a single observation.

For myself, I can say that this word "alway" was, from the first, the great difficulty with me—and the more I became convinced of the studied meaning with which Chaucer chose his other expressions, the less satisfied I was with this; and the more convinced I felt that the whole line had been corrupted.

In advocating the restoration of the reading which I have already suggested as the original meaning of Chaucer, I shall begin by establishing the *probability* of his having intended to mark the moon's place by associating her rising with that of a known fixed star—a method of noting phenomena frequently resorted to in ancient astronomy. For that purpose I shall point out another instance wherein Chaucer evidently intended an application of the same method for the purpose of indicating a particular position of the heavens; but first it must be noted, that in alluding to the Zodiac, he always refers *to the signs*, never to the constellations—in fact, he does not appear to recognise the latter at all!* Thus, in that palpable allusion to the precession of the equinoxes, in the Frankeleine's Tale—

> "He knew ful wel how fer Alnath was shove
> From the hed of thilke fixe Aries above."

—by the *hed of Aries*, Chaucer did not mean the os frontis of the Ram, whereon Alnath still shines conspicuously, but the equinoctial point, from which Alnath *was shove* by the extent of a whole sign.

---

* See Note B at the end.

This being premised, I return to the indication of a point in the ecliptic by the coincident rising of a star; and I contend that such was plainly Chaucer's intention in those lines of the Squire's Tale wherein King Cambuscan is described as rising from the feast:—

> "Phebus hath left the angle meridional,
> And yet ascending was the beste real,
> The gentle Leon, *with his Aldryan*"

Which means that *the sign* Leo was then in the horizon—the precise degree being marked by the coincident rising of the star Aldryan.

Speght's explanation of "Aldryan," in which he has been copied by Urry and Tyrwhitt, is—"a star in the neck of the Lion." What particular star he may have meant by this, does not appear: nor am I at present within reach of probable sources wherein his authority, if he had any, might be searched for and examined; but I have learned to feel such confidence in Chaucer's significance of description, that I have no hesitation in assuming, until authority for a contrary inference shall be produced, that by the star "Aldryan" he meant REGULUS, not the neck, but the heart of the lion—

1st. Because it is the most remarkable star in the sign Leo.

2nd. Because it was, in Chaucer's time, as it now is, nearly upon the line of the ecliptic.

3rd. Because its situation in longitude, about two-thirds in the sign Leo, just tallies with Chaucer's expression "*yet* ascending,"—that is, one-third of the sign was still below the horizon.

Let us examine how this interpretation consists with the other circumstances of the description. The feste-day of this Cambuscan was "The last idus of March"—that is, the 15th of March—"after the yere"—that is, after the *equinoctial year*, which had ended three or four days previously. Hence the sun was in three degrees of Aries—confirmed in Canace's expedition on the following morning, when he was "in the Ram foure degrees yronne," and his corresponding right ascension was twelve minutes. Now, by "the angle meridional" was meant the two hours *inequall* immediately succeeding noon (or while the "1st House" of the sun was passing the meridian), and these two hours may, so near the equinox, be taken as ordinary hours. Therefore, when "Phœbus hath left the angle meridionall," it was two o'clock P.M., or eight hours after sunrise, which, added to twelve minutes, produces eight hours twelve minutes, as the ascending point of the equinoctial. The ascending point of *the ecliptic* would consequently be twenty degrees in Leo, or within less than a degree of the actual place of the star Regulus, which in point of fact did rise on the 15th of March, in Chaucer's time, almost exactly at two in the afternoon.*

Such coincidences as these could not result from mere accident; and, whatever may have been Speght's authority for the location of Aldryan, I shall never believe that Chaucer would refer to an inferior star when the great "Stella Regia" itself was in so remarkable a position for his purpose; assuming always, as a matter of course, that he referred his phenomena, not to the country or age wherein he laid the action of his tale, but to his own.

This, then, is the precedent by which I support the similar, and rather startling interpretation I propose of these obscure words "In mena Libra alway."

There are two twin stars, of the same magnitude, and not far apart, each of which bears the Arabic title of Min al auwâ; one (β Virginis) in the sign Virgo—the other (δ Virginis) in that of Libra.

The latter, in the south of England, in Chaucer's time, would rise a few minutes before the autumnal equinoctial point, and might be called *Libra* Min al auwâ either from that circumstance, or to distinguish it from its namesake in Virgo.

Now on the 18th of April this Libra Min al auwâ would rise in the neighbourhood of Canterbury at about half-past three in the afternoon, so that by four o'clock it would attain an altitude of about five degrees—not more than sufficient to render the moon, supposing it to have risen with the star, visible (by daylight) to the pilgrims "entrying at a towne's end."

It is very remarkable that the only year, perhaps in the whole of Chaucer's lifetime, in which the moon could have arisen with this star on the 18th of April, should be the identical year to which Tyrwhitt, *reasoning from historical evidence alone*, would fain attribute the writing of the *Canterbury Tales*. (Vide Introductory Discourse, note 3.)

On the 18th of April, 1388, Libra Min al auwâ, and the moon, rose together about half-past three P.M. in the neighbourhood of Canterbury; and Tyrwhitt, alluding to the writing of the *Canterbury Tales*, "*could hardly suppose it was much advanced before* 1389!"

---

*See note C, at the end.

Such a coincidence is more than remarkable—it is convincing: especially when we add to it that 1388 is the very date that, by a slight and probable injury to the last figure, might become the *traditional* one of 1383!

Should my view, therefore, of the true reading of this passage in Chaucer be correct, it becomes of infinitely greater interest and importance than a mere literal emendation, because it supplies that which has always been supposed wanting to the *Canterbury Tales*, viz., some means of identifying the year to which their action ought to be attributed. Hitherto, so unlikely has it appeared that Chaucer, who so amply furnishes materials for the minor branches of the date, should leave the year unnoted, that it has been accounted for in the supposition that he reserved it for the unfinished portion of his performance. But if we consider the ingenious though somewhat tortuous methods resorted to by him to convey some of the other data, it is by no means improbable that he might really have devised this circumstance of the moon's rising as a means of at least *corroborating* a date that he might intend to record afterwards in more direct terms.

I acknowledge that, from the first, if I could have discovered a a probable interpretation of "mene" as an independent word, I should have preferred it rather than that of making it a part of the Arabic name, because I think that the star is sufficiently identified by the latter portion of its name, "Al auwâ," and because the preservation of "mene" in its proper place in the line would afford a reading much less forced than that I was obliged to have recourse to. Perhaps some Arabic scholar will explain the name "Min al auwâ," and shew in what way the absence of the prefix "Min" would affect it.*

[*Pub. May* 31, 1851.]

V.

TESTS OF POSITIONS.

As a conclusion to my investigation of this subject, I wish to place upon record the astronomical results on which I have relied in the course of my observations, in order that their correctness

*See Note D, at the end.

may be open to challenge, and that each reader may compare the actual phenomena rigidly ascertained, with the several approximations arrived at by Chaucer.

And when it is recollected that some at least of the facts recorded by him must have been theoretical—incapable at the time of actual observation—it must be admitted that his near approach to truth is remarkable; not the less so that his ideas on some points were certainly erroneous; as, for example, his adoption, in the *Treatise on the Astrolabe*, of Ptolemy's obliquity of the ecliptic in preference to the more correct value assigned to it by the Arabians of the middle ages.

Assuming that the true date intended by Chaucer was *Saturday, the 18th of April*, 1388, the following particulars of that day are those which have reference to his description:—

		H.	M.
Right Ascension	Of the Sun at noon	2	17·2
	Of the Moon at 4 p.m.	12	5·7
	Of the Star (δ Virginis)	12	25
		°	′
North Declination	Of the Sun at noon	13	47·5
	Of the Moon at 4 p.m.	4	49·8
	Of the Star (δ Virginis	6	43·3
Altitude	Of the Sun at 10 a.m.	45	15
	Of the Sun at 4 p.m.	29	15
	Of the Moon at 4 p.m.	4	53
	Of the Star at 4 p.m.	4	20
Azimuth	Of the Sun at rising	112	30
		H.	M.
Apparent Time	Of the Sun at half Azimuth	9	17 a.m.
	Of the Sun at Altitude of 45°	9	58 a.m.
	Of the Sun at Altitude of 29°	4	2 p.m.
	Of Moon's entrance to Libra	3	45 p.m.

[*Pub. June* 28, 1851.]

NOTE A.

"—— and the yonge sonne
Hath in the Ram his halfe cours y-ronne."—Page 67.

I am in a manner compelled to invite a comparison between this reprint of what was written and published by me so far back as 1851, and a claim advanced on the part of the Rev. W. W. Skeat, of Cambridge—that he, in 1868, was the first to protest against Tyrwhitt's proposal to read "*the Bull* instead of *the Ram*" in the above lines: for notwithstanding that Mr. Skeat did, at one time, publicly acknowledge that this claim of his had been groundless (Notes and Queries, 24th Oct., 1868), yet he afterwards permitted it to be reasserted, and apparently acquiesced in attempts to confer upon it great literary notoriety in connection with his name. It was made a prominent theme of congratulation for the members of the *Early English Text Society* in its Annual Report for 1869: and it was proclaimed, with a rather exuberant prefatory flourish, in the "Temporary Preface" sent out in May, 1869, with the first issue of publications by the *Chaucer Society*. But although I am thus driven to assert once more my long prior right to this correction of Tyrwhitt, it is not from any overweening pride in it, for I was not, I confess, at all impressed with its value, until enlightened by the high importance I now find attributed to it as the discovery of another person:—

"The greatest gain of late times as to the Prologue is clearly Mr. Skeat's showing that Chaucer's *Ram* of line 8 is not the blunder for *Bull* that Tyrwhitt and his followers supposed it to be; but is quite right."

"A vote of thanks to Mr. Skeat from all Chaucerians is hereby recorded."
("Temporary Preface," page 89.)

This is followed by a reprint in full of Mr. Skeat's letter to *Notes and Queries* of 19th September, 1868; but his disclaimer, which appeared in the same periodical a few weeks later, is not noticed. I fear Mr. Skeat himself cannot be wholly acquitted of having supervised this partial republishing of his letters: and I say so on the evidence of a superadded foot-note, not in the original, which nevertheless appears with the signature W. W. S. on page 104 of this "Temporary Preface;" and that foot-note clearly refers to an incidental remark in the remonstrance from me (N. and Q., 10th Oct., 1868), which had elicited his disclaimer on the 24th of the same month.

Now, while Mr. Skeat's protest against Tyrwhitt is identical with mine, the manner in which he would explain the existing text is essentially different; and it is necessary, in defence of my own interpretation, that I should show the errors of his. He would inflict upon Chaucer a more injurious imputation than even Tyrwhitt's—namely, an unintelligible jumbling together of the signs, of the Zodiac, with the constellations of the same names. He states that by the sun's halfe cours in the Ram Chaucer did not mean the *sign* so called but the *constellation;* and that since the middle of the constellation, in Chaucer's time, preceded the middle of the sign by twenty degrees, that number added to 15 of Aries would produce the 5th degree of Taurus, the sun's place on the 17th April. But, if that were Chaucer's meaning, what are we to think when we come to The Squire's Tale, and find there an almost exact repetition of the same phrase—"the yonge sonne that in the Ram was foure degrees y-ronne?" By parity, these four degrees should be advanced to twenty-four. But twenty-four degrees in Aries would be the sun's place for the sixth of April, whereas Chaucer declares it is the sixteenth of March (that is, the day succeeding "the last idus of March"). Therefore Mr. Skeat's interpretation would impute to Chaucer that by the same form of words he means a constellation in one place and a sign in the other, with a difference between the two meanings of twenty degrees of longitude!

Mr. Skeat makes no attempt to show by any authority other than "*a glance at a modern celestial globe*" what would have been Chaucer's estimate of the constellation Aries?—where he would place its beginning, where its middle,—or where its termination? The only allusion, I believe, in Chaucer's works to precessional separation, is the passage in the Frankelein's Tale quoted by me in 1851:—

> "He knewe ful wel how ferre Alnath was shove
> Fro the hed of thilke fixe Aries above."

and if any conclusion can be drawn from this, it is that he considered the star α Arietis to be the beginning of the constellation: therein differing from Ptolemy who almost excluded Alnath by ranking it with the outsiders. It is needless to say that Alnath is not the star with which the constellation begins upon "*a modern celestial globe.*"

The overlap of half a sign into each month, and reciprocally of half a month into each sign, which was *my* interpretation of the term "halfe cours," had no doubt become, in Chaucer's time, a conventional association arising from the still surviving habit of regarding the beginning of each sign as coincident with the middle of each month. It is so placed in the Treatise on the Chilindre translated and published by Mr. Brock; and it is so placed by Sacro Bosco, an English astronomer of the 13th Century, who wrote a small treatise "De Anni Ratione" in the year 1244:—

> "Si in quo gradu cujuslibet signi sit Sol scire volueris, numero dierum mensis præteritorum adde 15, et si resultent 30, vel minor numerus, in tali gradu signi ad mensem pertinentis est Sol," &c.

Now even this expression "*ad mensem pertinentis*" throws some light upon Chaucer's choosing to refer the sun's place in April to his completed passage through Aries rather than to his absolute presence in Taurus. Each sign was supposed to *belong* to that month in which it ended. Or as Sacro Bosco expresses it—

> "Nam signum detur mensi quem fine meretur."

I confess I cannot imagine a plainer or more completely *satisfying* explanation of the "halfe cours" in the Ram, than the overlap into April of the half sign Aries; through which it is said—(a few days after the middle of that month)—the sun *hath yronne*.

So obviously true does that interpretation appear to me, that I think it superfluous at present to support it by the further argument that Chaucer did not recognize the zodiacal constellations at all as apart from the signs. I shall, however, in another note, recur to that position; which, of itself, would be sufficient to refute Mr. Skeat's interpretation.

And although I have assumed, for the sake of argument, that Chaucer would consider Alnath to be the initial star of the constellation Aries, yet I am convinced that he was not alluding to the constellation, in the passage quoted from the Frankelein's Tale, but only to the individual star, as indicating by its removal from the equinoctial point the extent of precession. (On this subject see Introduction p. 14.)

It may be seen above, that in the same communication wherein I originally exposed the error of Tyrwhitt's "Bull," I pointed out another error, the misreading, in the standing text of the Canterbury Tales (10007), of *Juil* for *Juin*—on the ground that the sun could not be "in Geminis" on the 8th of *July* in Chaucer's time. By a singular coincidence Mr. Skeat also supplemented *his* communication on the same subject by a similar reference to the same line in The Marchaunt's Tale. He, however, would defend the error by asserting that the sun, though not in *the sign*, "would be in *the constellation*

Gemini—as so expressly stated by our poet." In this Mr. Skeat is unfortunately not correct. For the sun now leaves the constellation Gemini, as shown by "a modern celestial globe," on the 17th of July; which, when decreased by 14 (according to Mr. Skeat's own method of reduction to Chaucer's time), would result in the 3rd of July; several days before the day named in the present text of *The Marchaunt's Tale.*

NOTE B.

"In alluding to the zodiac he always refers to the signs, never to the constellations."—[p. 76.]

I intimated in the preceding note that I should recur to this assertion with the truth of which I am still, in 1869, perfectly satisfied. In fact the ignoring of the old misplaced and unequal constellations, in order that the names, fables, and attributes which had been associated with them in relation to the zodiac might be transferred to the equal divisions of the ecliptic called signs, seems to have become the practice long before Chaucer. And a much wiser and more sensible practice it was than the confusing double identities of signs and constellations which were reverted to in later times—solely, perhaps, to take advantage of the convenience presented by Ptolemy's constellations in classifying and cataloguing the fixed stars and comparing their positions with those assigned to them by him. Sacro Bosco describes the zodiac almost in the same words afterwards used by Chaucer (*supra* p. 30).

"Zodiacus a ζωη, quod est vita, quia secundum motum planetarum sub illo est omnis vita in rebus inferioribus: vel dicitur a ζωδιον, quod est animal, quia cum dividatur in duodecim partes æquales quælibet pars appellatur signum, et nomen habet speciale a nomine alicujus animalis, propter proprietatem aliquam convenientem tam ipsi quam animali: vel propter dispositionem stellarum fixarum in illis partibus admodum hujusmodi animalium."—

This description clearly ignores any groups of zodiacal stars other than those in the signs.

The fictions of animal configuration were originally typical of the seasons; and when in course of time, the seasons had moved away amongst the stars, it would doubtless appear only reasonable that those animal identities should move away with them and become associated with new stars. What else could Sacro Bosco mean by explaining the animal nomenclature of the signs by "*propter dispositionem stellarum in illis partibus*"—*i.e.*, in the *duo-im partes æquales* he had just described, and of which he says—

"*quælibet pars appellatur signum?*" And what else could Chaucer mean when, in like manner explaining why *the signs* should have names of *bestes*, he repeats the same description almost verbatim—"or else for that the stars *that there ben fixe* ben *disposed* in figure of bestes or schape like bestes"? This assigning of groups of stars to the *signs* is wholly incompatible with any idea of other groups of stars as constellations.

In the Teseide of Boccaccio, from which Chaucer took so much of his Knight's Tale, the following stanzas very aptly exemplify the entire transfer of the animal identity from the constellation to the sign:—

> Febo salendo cogli suoi cavalli,
> Dal ciel teneva l'umile animale
> Che Europa portó sanza intervalli
> Là dove il nome suo dimora avale;
> E con lui insieme graziosi stalli
> Venus facea de' passi con che sale
> Perchè luceva il cielo tutto quanto:
> E Ammon con Pesce dimorava intanto.
>
> Da questa lieta vista delle stelle
> Prendie la terra graziosi effetti,—
> Libro terzo 5. 6.

Here the sign Taurus is not only an animal, but the very "*umile animale*" that bore off the daughter of Agenor, she:—

> "That made great Jove to humble him to her hand
> When with his knees he kist the Cretan strand."

[It is worthy of note that Shakespeare and Boccaccio should both use the same expression *humble*: But surely Shakespeare must have written *Tyrian* and not *Cretan* strand. When Jove landed in Crete he was no longer *humble*.]

Boccaccio's description is *astrological*. The Sun and Venus in conjunction in Taurus, and Jupiter *sextile* in Pisces: constituting the *lieta vista*, or happy aspect of the stars. This proves that it is the sign is referred to and not the constellation, since Astrology deals only with the signs. And if further proof were needful it would be found in the allusion to the *graziosi stalli*, or fair halls of Venus, her *mansion* being the sign Taurus.

Chaucer has a somewhat similar allusion to the sign Taurus as the Cretan Bull, in the Prologue to his "Legende of Goode Women;" where, on the "firste morwe of May" he speaks of the sun—

> "That in the brest was of the beste that day
> That Agenores doghtre ladde away."—113-14.

By the breast of Taurus he could not mean the star so called by Ptolemy, which would be four or five degrees in advance of the sun's place on the 1st of May. He probably meant it as a general term for the Pleiades, the cosmical rising of which was of old the harbinger of summer—since the advent of summer is his theme:—

> "Welcome somer oure governour and lorde."—170.

Again, in Chaucer's poem, "The Complaynt of Mars and Venus," he allegorically describes a conjunction of the Sun with Venus and Mars, in Taurus. Venus had made an assignation with Mars in her "nexte paleys"—*i.e.*, the sign Taurus, as mentioned above. Her chamber "depeynted was with white boles grete,"—emblematic of Taurus—in which, as in the old fable, the Sun surprises her with Mars, by entering into Taurus—"*thys twelve dayes of Avrille*"—a date that of itself is sufficient to prove that it is the *sign* Taurus is alluded to. The adjoining sign to Taurus is Gemini, and Gemini is a mansion of Mercury just as Taurus is of Venus. It is needless to say that Mercury is *Cyllenius*, and when Phœbus so rudely bursts into Venus' chamber she escapes into Mercury's:

> "Now fleeth Venus into Cyllenius tour
> With voide cours, for fere of Phebus lyght."

Or, as Boccaccio hath it, "Perchè luceva il cielo tutto quanto." It was by the unravelling of this little astronomical allegory that I was enabled to declare, in 1851, that *Cyllenius* is the proper and obvious reading of "Ciclinius," an unintelligible name with which the above two lines are always printed—although I am not aware that the correction has been as yet adopted by any editor of Chaucer's poetical works.

NOTE C.

> "Phebus hath left the angle meridional
> And yet ascending was the beste real
> The gentle Lion."—Page 77.

If it could be known with any certainty what Chaucer meant by the *angle meridional* in this place, there would be no difficulty in determining at what time the sun would leave it on any given day and what the ascendant would be at that moment.

In technical astrology the angle meridional is *the tenth house*, the beginning of which is that point of the ecliptic then on the meridian, as Chaucer himself teaches in his XXXIXth problem (*ante*, page 55). If this were the meaning of the expression in the present instance the sun would have left the angle meridional the moment he had passed the meridian or point of noon: and the coincident ascending point of the ecliptic (on the 15th March, in Chaucer's time, and with his latitude and obliquity) would be almost exactly in the beginning of the sign Leo.

But this interpretation is open to two objections:—

First, the words "yet ascending was the beste reale," would imply that the sign Leo had been already ascending for some indefinite time: and if "yet ascending" is a true reading, it is not easy to see how any other meaning than previous continuance can be given to it.

Second, twelve o'clock at noon seems too early in the day for a royal birth-day banquet to be *concluded*; especially one so elaborately prepared as King Cambuscan's—

> Of which if I shal tellen al the array
> Than wold it occupye a somers day.

A feast of many courses—unusually prolonged by the interlude of the knight on the brazen horse—who had his speech to recite, his offerings to present, and his toilette to make ere he could join the banquet, which was not finally concluded 'till the king "rose from the bord."

All this seems inconsistent with noon. For assuming that the ordinary dinner time was at prime, which I shall elsewhere show to be nine o'clock A.M., *set-feasts* of even minor ceremony were generally later. As, for example, that family entertainment to which Creseide was entrapped by Pandarus, which, without any assigned reason

> ——"the faire Queen Heleine
> Shope her to ben an hour after the pryme."—T. C. ii., 1557.

It was the consideration of these two objections that induced me, in 1851, to adopt two o'clock P.M., when Regulus was on the horizon, as the break up of King Cambuscan's feast. And I should still hold to that opinion if it were in my power to justify by any technical reference the explanation I then gave of *Angle Meridional.*

But, as I cannot do that, I am free to acknowledge that the twelve o'clock hypothesis, based upon the absolute astrological meaning of angle meridional, now appears to me to be the unavoidable interpretation. And I am the more inclined to believe it to be the true one from finding a star in the ascendant quite as favourably situated for twelve o'clock as Regulus was for two.

That star is the southern Asellus (δ Cancri) which, like Regulus, was and is almost on the line of the ecliptic, and it would rise with the first degree of the sign Leo at two or three minutes past noon on the day in question, the 15th of March about the year 1390.

The only way at present in which I can reconcile this star with the name *Aldryan*, which purports to be the name given by Chaucer as that of the coincident ascendant, is by supposing that Aldryan may possibly be a substitution for *Hamaran*, which is the Arabic name of δ Cancri, and which would rhyme equally well with Cambuscan.

NOTE D.

" Perhaps some Arabic scholar," &c.—Page 79.

I am not aware that this invitation was ever responded to : but I have myself examined " Hyde's commentary on Ulugh Beigh," and therein I found the star δ Virginis with the Arabic name " min al away " attached to it, and translated by Dr. Hyde, " De latratore seu vociferatore," min being represented by the preposition *de*, which is no doubt its obvious and literal meaning. Nevertheless, being still convinced that some more significant explanation might be discovered, and observing that the same prefix *min* was also applied to four or five other stars, all of which are mansions of the moon, *and to no others;* I suspected that *min* must in some way have been a form or abbreviation of the Arabic *Menzil*, *mansio Lunæ.* And this opinion was strengthened by finding in Fretagh's Arabic Lexicon, iv. 214, " Mina, ex licentia poetica pro Manazil, *mansiones*."

No doubt this is a plural where we want a singular,—but I think it gives sufficient authority to assume that there might also have been an abbreviation of the singular *menzil.* At all events, there is another word *Mind*, the signification of which is a port or harbour for ships, a meaning not altogether remote from " a station or stage in a journey." And when the changes of signification that occur in course of time in all living languages are taken into account, the absence in modern lexicons of the precise definition of *mina* required for this reading is no very conclusive reason against it.

If such an explanation be admitted for "*min al away*," the passage in Chaucer might be read in this way:—

> —"then with the mone's exaltation
> In min al auway Libra gan ascend;"

that is, coincident with the moon's rising in her mansion al away, Libra gan ascend: being virtually the same position as that laid down by me in 1851, although explained on different, and, I think, on better ground.

Sir William Jones (Works, vol. i., 346) has the following:—

> "Menzil, or *the place of alighting*, properly signifies a station or stage, and hence is used for an ordinary day's journey; and that idea seems better applied than mansion to so incessant a traveller as the Moon; the menazilu'l kamar, or lunar stages of the Arabs, have twenty-eight names in the following order, the particle *al* being understood before every word."

It is unnecessary to follow Sir William Jones through the whole list of twenty-eight mansions; it is sufficient to observe that amongst them are:—

2nd Butain, ϵ Arietis,	called	Min Butain
11th Zubrah, θ Leonis,	"	Min al Zubra
13th Awwa, δ Virginis	"	Min al Auwa
20th Naiim, γ Sagittarii	"	Min al Naiim
22nd Dhabili, β Capricornis,	"	Min al Dabih

Now the frequent recurrence of this same prefix *Min* to so many of the manazil is surely a presumptive proof that it was a name special to that designation. If it simply signified the preposition *of*, it would have been applied to other stars or constellations in the firmament as well as to these.

And that Chaucer was well acquainted with these mansions sufficiently appears in The Frankelein's Tale, where the Astrologer's book

> "——— spoke moche of operacions
> Touching the eight and twentie mansions
> That longen to the Mone."

ON THE MEANING OF CHAUCER'S PRIME.

The various allusions in Chaucer's works to *prime* as a certain time in the day are apparently so contradictory, that it seems impossible to assign any single hour that shall satisfy them all. I am inclined to believe that in the common phraseology of the time there were at least two different periods of the day called prime—one in the forenoon, and another in the afternoon—and that both were considerably later in the day than the time usually attributed to prime in supposed conformity with canonical regulation. I do not pretend to assert that an earlier prime may not have been intended by Chaucer in any of the passages wherein it is mentioned: what I wish to say is, that I have met with none that may not equally well be reconciled with one or other of the primes I am about to assume.

What are called "canonical hours" are in themselves extremely vague and uncertain, often varying with the rules of the different religious orders. Some authorities declare that prime was at absolute sunrise; others at conventional sunrise, which was six o'clock all the year round; and others, again, at seven o'clock a.m., when the first hour was noted; for it seems a reasonable deduction that where noon was sexta, five hours before noon should be prima.

John de Belethus, who wrote on ritual observance about the end of the twelfth century, in his chapter, "Cur septies in die laudemus Dominum," likens the seven canonical hours of the day to the seven ages of human life: and it is to the *second* age (Shakespeare's school-boy) that he assigns prime:

> "Per matutinas laudes representatur infantia: per primam pueritia: per tertiam adolocentia: per sextam juventus: per nonam ætas virilis: per vesperas senectus: per completorium ætas decrepita ac finis humanæ vitæ."

But the same writer in his next chapter apportions the twelve ordinary hours as follows:—

> "Sub prima horas duas complectimur ipsam videlicet primam et secundam; sub tertia tres, ipsam tertiam et quartam et quintam: sub sexta itidem tres ipsam sextam septimam et octavam: vesperæ representant undecimam: completorium duodecimam."

Here are all the twelve hours accounted for, but only six out of the seven canonical divisions before enumerated. In appearance, nothing can be more distinct and specific, but in reality nothing can be more ambiguous. Where is *infantia?* did it precede sunrise, the birth of the day, or did decrepitude not begin until after dissolution at sunset? At which end of the day was the overlap? Were the several points of time, tertia, sexta, nona, &c., initial or terminal to the hour spaces of the same names? Did the three-hour division, comprising sexta, septima, and octava, begin at noon or at some time previous to noon? These are some of the difficulties that beset the subject of canonical hours, and render them so indeterminate. Much of the uncertainty is caused by the use of Latin ordinal numbers in a cardinal sense, which must always tend to ambiguity when the arbitrary meanings conventionally attaching to them have become lost or forgotten. It is the fate of the horary reckoning of the Romans themselves, leaving the true interpretation of Martial's distribution of hours and employments ever open to dispute.

It is fortunate that the meaning of Chaucer's prime may be determined without reference to the uncertain prime of the canonical hours. But before entering into its discussion it is necessary to eliminate the very erroneous notion recently introduced (see notes to the Preface of Mr. Brock's translated "*Treatise on the Chilindre,*" and to page 19 of Mr. Furnivall's "*Temporary Preface,*") that planetary, or unequal, hours were employed by Chaucer in his estimation of prime.

That he thoroughly understood such hours *in theory* is most certain,—as is shown in the way he introduces them into his story of Palamon and Arcite, as well as in the explanation he gives of them in the XIth and XIIIth Conclusions of his Astrolabe,—but it is equally certain that he neither practised them in ordinary life himself, nor attributed their practice to any time cœval with his own.

If ever he would do so it would most assuredly be in the Nun's Priest's Tale in the *natural* announcement of hours by Chanticleer, who knew the time "by kinde and by none other lore." But, so far from doing so, he indicates the contrary, not only by direct assertion, but by throwing in, with his usual love of collateral corroboration, a little scientific fact which places his intention beyond all possibility of cavil:—

"Wel sikerer was his crowyng in his logge
Than is a clok, or an abbay orologge,

By nature knew he ech ascencioun
Of equinoxial in thilke toun;
For whan degrees fyftene were ascendid
Thenne crewe he, it might not be amendid."

It is not so much the mention of clock or orologe in these lines, although that is significant enough, as the assigning to each hour fifteen degrees of the equinoctial that is so absolutely decisive of equal hours.

It is doubtful whether Chaucer would have so well understood even the theory of unequal hours if he had not made them an object of special study for the purpose of his Knight's Tale, into which the introduction of these hours was all his own, there being no allusion to them in Boccaccio's Teseide: for so completely were those hours forgotten long before his time that John Holywood (J. de Sacro Bosco), who was a regular professor of the sciences of which Chaucer was but an amateur, gives this grossly absurd mis-description of them in his libellus "De Anni Ratione" written in 1244:—

"Hora est vigesima-quarta pars diei naturalis. Horarum vero alia naturalis alia æquinoctialis. Naturalis est spacium temporis quo medietas signi peroritur. Æquinoctialis vero est 15 graduum circuli æquinoctialis supra horizontem ascensio."

That is to say, a natural or unequal hour is the ascension above the horizon of 15 degrees of the *ecliptic;* and an equal hour is the ascension of 15 degrees of the *equinoctial:* a very pretty antithesis but monstrously untrue. How untrue, is best seen in the fact that in April, when "*houres inequall*" ought to be *long,* half the sign Taurus would ascend in little more than half an ordinary hour—while in October, when those hours ought to be *short,* half the sign Scorpio would be nearly three times as long in ascending. Such an extraordinary misdescription shows that in the thirteenth century "houres inequall" had become so obsolete in common life that their nature was forgotten and misunderstood.

It may be said that Sacro Bosco in so describing the duration of what he called *hora naturalis* did not mean the hour that Chaucer calls "*houre inequall.*" Be it so: in that case his silence respecting such hours is quite as significant and proves equally well their desuetude in his time.

Tyrwhitt, with all his blunders of astronomical interpretation,*

* One of these has not, I think, been noticed. In the Merchant's Tale, line

did not attribute unequal hours to Chaucer's time; although he understood these hours well and fully explained them in two of the best notes in his whole series. Why they should be imputed now is not very apparent,—unless, perhaps, it is because Mr. Brock's *chilindre* is constructed for a twelve-hour division of the day. But that instrument has nothing in common with Chaucer except its name, and even that is mentioned by him only in metaphorical allusion, as I shall shew hereafter.

The several MS. descriptions of the *chelindrus*, one of which has been so well printed and translated by Mr. Brock, were probably copies, with slight variations, furnished from one convent to another, for the purpose of teaching the construction of an instrument to regulate their religious services.

The twelve-hour division of the day was long held in religious veneration on account of its supposed inculcation in St. John xi. 9, with reference to which Venerable Bede writes:

> ——"XII horæ diem complent, *Domino attestante*, qui ait, nonne duodecim horæ, etc. Ubi quamvis allegoricè se diem, discipulos horas, appellaverit?"

The "Abbey Orologe" would announce equal hours to the external world, but for religious matters some special guide to unequal hours would be all the more necessary from their total disuse in ordinary life.

I have now, I trust, cleared away the encumbrance of unequal hours from the discussion of Chaucer's prime,—and indeed it is a matter of surprise to me how any one could have associated them in the face of the plain declaration in The Nun's Priest's Tale, that prime occurred at one of Chanticleer's equinoctial hours;—in the month of May, too, when the lengthening of unequal hours would occasion a sensible difference in the reckoning. But of this announcement of prime by Chanticleer I shall have occasion to speak hereafter.

Of the two primes, mentioned by me at the commencement, I assume that the first was at nine o'clock, A.M., and that it was identical in time with what the Italians called *terza*. Of this identity

---

9761, Tyrwhitt altered "two of taure" into "ten of taure," *and so printed it;* explaining in a note to that line that he did so because the motion assigned to the moon in four complete days exceeded the *mean* motion in that time! As if Chaucer was necessarily confined to the mean motion: or, as if a true motion to the same extent, in the same time, as that stated by him, were not to be seen in every almanac.

the definition of that word in Florio's Italian Dictionary—"Terza, the third in order: also the hour that Priests call Prime."—is one presumption. Another arises from Boccaccio, in his Decameron, frequently referring to Terza the same incidents that Chaucer refers to Prime: thus terza is the dinner-time appointed for the personages of the Decameron just as prime is associated with dinner by Chaucer. Another point of resemblance is that the true time of terza seems to be as much a matter of uncertainty with the Italians as prime is with us. In their great national dictionary, Della Crusca, it is presented with the same unsatisfactory explanation—"*Una dell' ore canoniche*,"—word for word the explanation of prime by our own dictionaries. But in the old French translation of the Decameron by Antoine Le Maçon, which went through so many editions in the sixteenth century, terza is invariably rendered '*neuf heures*."

Nine o'clock in the morning may appear to us now a strange hour for dinner; but it is the name deceives us, for although called dinner, the meal was in truth breakfast—breakfast of that substantial kind "*a la fourchette*." In the Promptorium "Dynner" is defined as *jeutaculum* as well as *prandium;* and Du Cange tells us (in v. dejejunare) that "disner" is but an abbreviation of "desjeuner."

Another feature of identity between Boccaccio's terza and Chaucer's prime, is the frequent use by the former of the phrase *mezza terza*, which appears to be an exact counterpart of Chaucer's "half way prime." And here again we are met by the same uncertainty of meaning. The Della Crusca entirely ignores the phrase; and Le Maçon, so invariable in his rendering of *terza*, is all abroad in respect of *mezza terza*. Under the same circumstances he gives it several different interpretations—"Sur les sept ou huict heures"—"Entre six et sept heures"—"Entre sept et huict heures,"—&c. One point, however, of absolute certainty may be gathered from Boccaccio's context—that mezza terza was *antecedent* to terza in time.

Tyrwhitt, in his note upon "half-way prime," suggests that it was half-past seven o'clock, in which he was probably right; but then he supposed prime to be at six o'clock A.M., and consequently the earlier. He cites a passage from the "Modus tenendi Parliamentum," in which it is stated that Parliament was to assemble on ordinary days at "hora mediæ primæ" but on festivals at "hora prima," on account of divine service, Hence Tyrwhitt would understand that Parliament was to assemble *earlier* on festival days, whereas, in my opinion, the obvious meaning of the rule is that the sitting was to be *postponed* on festivals until after divine service: which it would be if prime be understood as nine o'clock, in complete analogy with the Italian phrases.

I have now to consider Dan Johan's prime in "The Shipman's Tale," where one thing at least is certain, that he associates it with "dinner," that is, nine o'clock, A.M. But then he says that it is prime *by his chilindre*, which some people understand as the result of absolute instrumental observation of the time *then present*. But *that* it could not be; because the time then present was considerably earlier than dinner-time, to permit of its preparation, and of the intervening celebration of mass. It is therefore infinitely easier to understand the mention of the chilindre as purely metaphorical. It is just like one of Chaucer's humorous touches to imagine the monk alluding to his own cylindrical casing, and calling it his chilindre:

> ——" let us dine as sone as ever you maye
> For by my chilindre it is pryme of daye."

There is a counterpart, though a very poor one, of the same joke in Middleton's play of "The Changeling:"—

> "What hour is it, Lollio?"
> "Towards belly-hour, Sir."
> "Dinner-time, thou meanst, twelve o'clock."

That the metaphorical interpretation was that which was formerly put upon Dan Johan's chilindre is obvious from the substitution of "stomach" quoted by Mr. Morris as "the reading of one MS."

The chilindre, or cylinder, seems to have been a very common instrument for several centuries before and after Chaucer's time. There is an engraving of one almost identical with the figure prefixed to Mr. Brock's translation (except that it represents an instrument some six centuries later in time and for the latitude of Paris) in the French "Encyclopedie Methodique," where it is entitled "Cylindre montè et pieces qui le composent:" and in the description of the same instrument by Dom Bedos it is called "Le Cylindre Portatif."

It is evident, therefore, that the name "chelindrus" was not particularly special to this instrument, but merely a variation in spelling. In the words quoted by Mr. Brock from Stœffler—"*velut est umbra stili in pariete aut chilindro*"—chilindrus is not more special than paries—the distinction is merely between the cylindrical surface of a pillar, on which a dial was often drawn, and a flat wall. There is even "*chelindrus* vide *cylindrus*" in Francis Holyoke's Dictionary, with no other speciality than a garden roller.

Great praise is undoubtedly due to Mr. Brock for having by his reprint and translation of the Treatise on the Chilindre brought into notice such a capital gloss of Dan Johan's expression; but in no other sense can the instrument it describes be considered as specially associated with Chaucer. Its construction is for a time at least two centuries earlier than his—when the beginning of each sign was coincident with the middle of each month. And yet in the following note to page 47 of Mr. Brock's translation, that very peculiarity, indicative as it is of the age of the instrument's construction, seems to be imputed to its rudeness and imperfection:—

> "In Chaucer's time Aries rose on the *twelfth* of March, not on the fifteenth, and similarly for other signs. Hence arises an inaccuracy in the use of the cylinder."

The last conclusion is not very clearly expressed, but it probably means that an inaccuracy would arise if the instrument were used in Chaucer's time. Of course there would—the same sort of inaccuracy that would arise from the use, this year, of last year's almanac. But, since the instrument was *not* constructed for Chaucer's time it seems scarcely a matter for remark that it does not agree with it. In another note on the last page an erroneous latitude and obliquity of ecliptic are attributed to the instrument, and that, too, in a conjectural sort of a way, as if any doubt ought to exist as to these elements when they are so very plainly indicated in the description. The solstitial altitudes, upper and lower, being given, half their sum is the co-latitude and half their difference is the obliquity. These altitudes are 61° 34′ and 14° 26′ respectively, so that the latitude is 52 degrees, and the obliquity 23°·34′ And if this were not a sufficient indication, the *equinoctial altitude* is stated in another place to be 38 degrees, equal, of course, to the co-latitude.

But return to the subject of prime. Enough has been probably said respecting the forenoon prime at nine o'clock A.M.: and as to that in the afternoon, I assume that it was at one P.M. And I rest this assumption,—

Firstly, upon the probability that, when the initial of horary notation was transferred from sunrise to noon, the term prime would from analogy be transferred with it; or, in other words, that where hour *one* was reckoned *prima* would be considered applicable.

Secondly, upon the excellent interpretation which the supposition of prime at one P.M. confers upon two passages of Chaucer's text that cannot otherwise be so well explained. The first is that passage in Troilus and Creseide. Book v., 472:—

> "The letters eek which she in olde time
> Had him y-sent, he would alone rede
> An hundred sithe atwixe noon and prime."

The inconsolable Troilus—

> "When he was ther as no man might him here,"

is bemoaning in secret the loss of his mistress: but he is visiting amongst strangers in Sarpedon's house: what better time, then, could he choose to read over her old letters *alone*, than the hour of noon-tide siesta *between twelve and one*?

The second is that passage in The Squiere's Tale where the narrator exclaims—at a time which must have been in the afternoon—

> "I wol not tarien you for it is prime."

I am aware that this line has been otherwise explained by the assumption that the journey to Canterbury could not have been performed by the pilgrims in a single day, and that The Squiere's Tale might be related at an early hour of some other day. But to that hypothesis I cannot subscribe. Not only I disbelieve that such important incidents as haltings at nights, and assemblings in the mornings, could have been intended by the author, and yet not the slightest allusion be made to them; but I should esteem any attempt to test, by serious investigation on the mere score of distance, the practical possibility of performing the pilgrimage to Canterbury in one day, or in two, or in three, to be about as wise as a similar inquiry would be into the stages made by Imogen in her ride to Milford Haven—and about as likely to lead to a satisfactory determination.

I have said that I am not aware of any mention of prime by Chaucer that may not be referred to one or other of those I have assumed. There is, however, one mention of it in The Nun's Priest's Tale to which I have before alluded, and which it is more especially necessary to discuss, because, although at first sight it seems to require an earlier hour than nine o'clock A.M., it turns out, when properly investigated, to be the strongest possible corroborative of that hour—and because it tends to settle an uncertain text by showing the incorrectness of a reading first introduced by Urry, continued by Tyrwhitt, and repeated by every modern editor of the Canterbury Tales in preference to one which, though rejected by Twyrwhitt himself, was declared by him to be "the reading of the greatest part of the MSS." The month and day on which Chanticleer's catastrophe befell is of no further importance at present than as corroborating the longitude of the sun in Taurus; which is, say, 21½ degrees, equivalent to 18½ degrees of North Declination. And with that declination, what we are concerned with is the altitude of the sun, as indicating the hour of prime, assuming the latitude to be that of Chaucer's Astrolabe.

This altitude, as printed in the editions of Urry, Tyrwhitt, Wright, and Morris, is :—

> "Twenty degrees and oon and more i-wis,"

and to this line Tyrwhitt appends the following note—

> "The reading of the greatest part of the MSS., is fourty degrees. But this is evidently wrong; for Chaucer is speaking of the altitude of the sun at, or about, prime—*i.e.* six o'clock A.M. See ver. 15203. When the sun is in 22° of Taurus, he is 21° high about three-quarters after six A.M."

Here one scarcely knows which to admire most—that Tyrwhitt should consider three-quarters of an hour a sufficiently near approach for Chaucer to make to an intended astronomical position—or that he (Tyrwhitt) should assume "six o'clock A.M." for prime when seven would have been so much nearer to the result of his own calculation!

But neither six o'clock, nor seven—neither three-quarters, nor one quarter of an hour—will satisfy those who will follow and are capable of understanding the precision with which Chaucer has determined similar problems in the IVth Conclusion of his Treatise on the Astrolabe.

In the present case the latitude and obliquity of Chaucer's instrument being known, the sun's place in the ecliptic, together with his altitude, are given to find the hour of prime: the sun's place being further confirmed by the mention of the month and day.

There can be no doubt that Chaucer intended, by all these precise elements, to make a little display of his astronomical knowledge; and, therefore, *when there is a choice of reading*, it is clearly due to his text that preference should be given to that reading from which the nearest approximation may be obtained to the same degree of correctness he has shown elsewhere. It was this consideration that induced me to examine the result of a calculation at the higher altitude of the alternate reading—

"Fourty degrees and oon, and more i-wis,"

and I had the satisfaction to find a resulting hour for prime of *nine o'clock* A.M. *almost to the minute.*

After this, there surely can be no reasonable doubt that "fourty," and not "twenty," is, on its own merits, the true reading of the first word in this line; even if it were not corroborated by Tyrwhitt's acknowledgment that it is "the reading of the greatest part of the MSS.:" and even if it had not been the reading of all the printed editions anterior to Urry's.

The false reading in altitude was probably introduced by echo from the line above; or by the officious alteration of some early scribe who, like Tyrwhitt, might fancy that, at prime, the sun must necessarily be at a low altitude, as if the line—

"Cast up his eyes to the brighte sonne,"

did not, of itself, bespeak a *high* altitude.

I had already written these remarks before I became aware that the reading I am contending for had been strenuously advocated by Mr. Thomas Thynne, son of the first editor ef the collected works of Chaucer in 1532, and himself the possesser of many MSS. His "*Animadversions upon Chaucer's Works*" have been edited by Dr. Kingsley, and printed by the Early English Text Society, in 1865.

Francis Thynne was apparently unconscious of the powerful support his reading would receive from the astronomical position of the sun in connexion with prime. But it must be observed that the object of *his* correction was not to increase the altitude but to decrease the number of degrees in Taurus which then stood at *fourty-one* in all editions (his father's included) anterior to his "Animadversions," written in 1599: for it is a singular fact that, whereas in modern editions both quantities are printed *twenty-one,* in Thynne's time both quantities were printed *fourty-one.* And it is also singular that whereas I have attributed the modern repetition to "echo from the quantity above," Mr. Thynne should have attributed the repetition in his time to a somewhat similar cause:—

"But although there be no misnaminge of the signe, yet yt is true the degrees of the signe are misreckoned ; the error whereof grewe because the degree of the signe is made equall with the degree of the sonne ascended above the Horizon."

He then proceeds :—

"But to remedye all this, and to correct yt accordinge as Chaucer sett yt downe in myne and other written copies; and that yt may stande with all mathematicall proportione, whiche Chaucer knewe and observed there, the print must be corrected after those written copies (whiche I yet holde for sounde 'till I maye disprove them) having these wordes :

"when that the month in whiche the worlde beganne,
that hight Marche, when god first made manne,
was complete, and passed were also
since marche begonne thirty dayes and two
befell that Chanteclere in all his pride
his seven wives walkinge him beside
cast vp his eyen to the bright sonne
that in the signe of Taurūs had yronne
Twentye degrees and one and somewhat moore
And knewe by kynde and by noone other loore
That yt was pryme, and crewe with blisful steven
The sonne, quoth he, is clomben vp on heaven
Fortye degrees and one, and moore, y wis," &c.

This extract is precisely copied as it appears in the reprint; and I quote it because it just comprises the passage which I might have cited (with somewhat different spelling and arrangement) to illustrate my own argument. It ought to be noted that in the fourth line "begonne" is used by Thynne in the sense of *passed by*, as a conjectural emendation of "beganne." Had he suggested *had gone* instead of *begone* the emendation would, I think, be better and more probable. His argument—that any further time added to a month *complete*, with the word "*also*," must necessarily be applied at the end of the complete month and not at its beginning—is excellent, and, I think, unanswerable. Suppose, for example, the time to be added is *two* days, expressed in this way :—When the month called March was complete, and *also* two days since March began,—it would be stark nonsense. And, if that be true for two days, it is not less true for thirty-two or sixty-two. The point from which *also* must start, is that previously arrived at,—that is, March complete. Having declared March to be complete we cannot undo it again and go back to its commencement. Such is the substance of Francis Thynne's argument.

Tyrwhitt saw this when he framed the line :—

> "Sithen March ended, thritty dayes and two."

But Francis Thynne's suggestion (slightly varied) is nearer to the original and retains "since :"—

> "Since Marche had gone thirty dayes and two."

Mr. Morris adopts :—

> "Since March beganne twa monthes and dayes two."

which, though it amounts to the same thing, is nothing but a compromise to retain "beganne" at the expense of direct contradiction to "*also.*"

But I have not as yet exhausted all the evidence in support of Chaunteclere's prime being Nine o'clock A.M. The "Colefox ful of sleigh iniquitee."

> "The same nighte thurgh the hegge brast
> Into the yard, ther Chaunteclere the faire
> Was wont, and eek his wives, to repaire,
> And in a bed of wortes stille he lay
> Til it was passed undern of the day—
> Waiting his time on Chaunteclere to falle."

Now the acknowledged gloss of *undern* is nine o'clock A.M., and just before the catastrophe Chanticleer had announced prime. Therefore, either the story is not consistent, or prime and undern were synchronical. But if prime was nine o'clock A.M. then must the sun's altitude at prime on the 2nd May in Chaucer's time be

> Fourty degrees and oon and more y-wis.

THE CARRENARE.

I have lately seen a very extraordinary interpretation of a passage in Chaucer's "Booke of the Duchesse" which more particularly attracted my notice because I have long had a notion of my own respecting the same passage.

It is that in which the well known crux occurs—"Go hoodless into the drie see and come home by the Carrenare," and it occurs in a sort of monody spoken by a disconsolate knight upon his lost mistress, who would not, he says, practise the heartless wiles resorted to by others of her sex to attract admirers and display the power of their charms:—

> "Hyr lust to holde no wight in honde
> Ne, be thou siker, she wolde not fonde
> To holde no wight in balaunce
> By halfe word ne by countenaunce;
> Ne sende men into Walekye
> To Pruise ne into Tartarye
> To Alisaundre ne into Turkye
> And bid him faste: anone that he
> Go hoodless into the drye see
> And come home by the Carrenare."
>
> 1018—1028.

Upon this, the author of a book, published last year, called "Chaucer's England" comments as follows:—

> "The last three lines are banter: *q. d.* 'nor send him to fetch her a pound of green cheese from the moon."——

the lines so described being spoken by a knight whose grief for the loss of his mistress is so overwhelming that—

> "It was great wonder that nature
> Might suffre any creature
> To have such sorwe and be not ded."

But the comment continues:—

> "It may perhaps be for want of vision, but I confess that I see no obscurity here. Of course the 'dry sea' is an absurdity, it was meant to be so. As for the word *carrenare*, it is a stumbling-block, but not a worse stumbling-block than some other adapted, modified, or mangled words in Chaucer. I take it to be bad Italian for carrier or caravan. If we suppose the word to have been written *carrattare*, it is scarcely very bad Italian. The proper word would be *carrettiere* a carter; but *carretta* means cart, and *carrettare*, formed from that for the sake of the rhyme, is not very outrageous license, compared with other things of the same kind to be found in Chaucer and poets of the time."
>
> *Chaucer's England*, Vol. 1. p. 62.

I have quoted this curious comment at full length from the impossibility of doing it justice with less, and because it affords a fair specimen of the sort of criticism to which this word has been exposed. We are to suppose that the lady is eulogized for not being disposed to order her knight to come home by the caravan, or, as she might say in these days, by the omnibus: but it does not appear why it should be a virtue to refrain from that very harmless injunction; and it is equally difficult to understand how the "scarcely very bad Italian" word *carrettare* can have been adopted "for the sake of the rhyme" when the rhyme required is to the English word *ware*.

Now, independently of the grave nature of the subject, the banter and green-cheese point of view is the last from which these injunctions should be regarded, when it is recollected how prone some of the dames of chivalry were to exact from their "*Servants*" cruel and dangerous proofs of their valour and devotion. As, for example, that fair lady who dropped her glove into the arena before a raging lion, in order that her knight, the famous Don Miguel Ponce *de Leon* (for he obtained the surname by the exploit) might descend and recover it.

The beautiful and amiable lady belauded in Chaucer's dream was of a different nature—"She ne used no such knackés smale"—but if the ladies with whom she is contrasted are not spoken of in earnest where would be the force of the comparison? The parting injunction shows a full determination that it was to be no joke:—

> "And, Sir, be now right ware
> That I may of you here saien
> Worshippe, or that ye come ageyn."

Moreover, the dangers to be encountered by the wight are in keeping with the known wars of the time. Pruise, where the Teutonic knights were waging a savage and sanguinary crusade—Walakye—Tartarie—Turkye—Alisaundre! And it is significant of the real nature of the enterprises that they are nearly the same as those attributed to THE KNIGHT in the Prologue to the Canterbury Tales.

Now my interpretation of the Carenare is that it is the gulf of the Carnaro in the Adriatic.

Il Carnaro, the charnel-hole: so called because of its reputed destructiveness of human life.

Chaucer's residence in Italy would make him well acquainted with the character of this gulf (now called *Il Quarnero*): and if it is true that he visited Padua, he would have been in the very place to hear of it. It is, indeed, from a Paduan writer, Palladio Negro, that the Abbé Fortis quotes—"E regione Istriæ, sinu Palatico, quem nautæ carnarium vocitant," &c., showing, by this translation of the name into the Latin equivalent *carnarium*, that Carnaro was not merely a name, but a nickname expressive of its fatal reputation.*

But the most conclusive description is by Vergier, Bishop of Capo d'Istria, as quoted by Sebastian Munster in his "Cosmographie," page 1044 (Basle Edition).

> "Par deça le gouffre enragé lequel on appelle vulgairement Carnarie, d'autantque le plus souvent on le voit agitè de tempestes horribles ; et là s'englontissent beaucoup de navires et se perdent plusieurs hommes."

If it be objected that carnaro is not carrenare, it is an objection that might be shown in many ways to be of no moment. The shortest answer is perhaps this,—that if Palladio Negro might translate the epithet into Latin, so might Chaucer into English from his own "Careyn"—careynare—to rhyme with *ware* in the line following.

It may be that Chaucer was reminded of the fatal character of the Carnaro by Dante's allusion to it in the Inferno (ix. 112):

> " ——— a Pola presso del carnaro
> Fanno i sepolcri tutto il luogo varo"—

which at all events shows that it was at that time a byword of danger and destruction.

It is a singular coincidence that on the same page of Munster wherein he quotes the description of the Carnaro there should be an account of a neighbouring inland lake, intermittent every half year, which might, in its dry season, be the "drie sea" coupled with the Carrenare by Chaucer. I do not, however, suggest it with anything like the same confidence that I do the Carnaro: because "drie sea" is a description so wide and uncertain, and is consequently open to so many different interpretations, that unless some special reference should be discovered to throw light upon it it is scarcely capable of more than the loosest suggestion. The following is a translation of Munster's description:—

> "It is said that there is a lake near the city of Labac, adjoining the plain of Zircknitz, which in winter time becomes of great extent, and abounds in fish of great size, that are taken with spears two or three ells in length. But in summer the water drains away—the fish expire—the bed of the lake is ploughed up—corn grows to maturity—and, after the harvest is over, the

---

* We, too, on the coast of England have our "*Shambles*"—a dangerous shoal off the Bill of Portland being so called.

waters return with the approach of winter, the lake is again filled, and the fish reappear. The Augspourg merchants have assured me of this, and it has been since confirmed to me by Vergier, the Bishop of Cappodistria."

Now it must be recollected that, although this account was written a century and a half after Chaucer, yet since this lake, according to a modern observation, published only the other day, and alluded to in my Introduction, is very nearly in the same state now, after the lapse of twice that time, it is a fair presumption that it also existed long before Chaucer: and there is no knowing what marvellous tales respecting it, and of the peculiar danger of traversing it *hoodless*, may have been popularly current in Italy when Chaucer was there. But then the difficulty is that the same might be said of any arid sandy desert that might be metaphorically called a dry sea. Consequently I only suggest this account of the lake at Zircknitz for what it may be worth, with whatever presumptive support it may derive from its proximity to the Carnaro. In the modern account the lake is stated to be ten leagues in length and one in breadth. A journey through it would necessarily be unwholesome; and as there must be an absence of all shade in the dry bed of an intermittent lake, danger from sunstroke might arise from going into it *hoodless*.

There is another possible interpretation of "drie sea." A *frozen* sea might be so called. And from a passage in Warton's History of English Poetry, Vol. I., page 461, it seems that to encounter severe cold *hoodless* was a feat in amatory chivalry. "It was a crime to wear fur on a day of the most piercing cold, or to appear with a *hood*, cloak, gloves, or muff."

SHIPPES OPPOSTERES.

As my interpretation of this expression (Knight's Tale, 2017) was published in the Athenæum newspaper in 1867 without my concurrence, and consequently without that support it might have received from my own explanation and advocacy, I take this opportunity to republish it, together with some of the reasons which have induced me to abide by it.

My interpretation is that the meaning of opposter is—opposer =

opposite = antagonist; and that the feminine plural form of *opposteres* was given to it by Chaucer in order to render it an absolutely literal representative of the *bellatrices* of Statius and the *bellatrici* of Boccaccio.

Such an interpretation would establish a remarkable consistency of expression in all three versions: each being compounded with a qualifying noun substantive used adjectively in its feminine plural form:—

Carinæ bellatrices ... Statius, Threbaid, vii. 57
Navi bellatrici......... Boccaccio, Teseide, vii. 37
Shippes opposteres ... Chaucer, Knight's Tale, 2017

It has, I believe, been objected that the formation of this word opposteres from the verb oppose, with the feminine Anglo-Saxon *estres* would not be consistent with the strict rules of composition. But if the word is in every other respect exactly suitable, are we to bind over Chaucer, in his word-craft, to strict conformity with our notions of proper composition? Will such objectors explain the formation of the word *divinistre*, in line 2811 of the same Knight's Tale?

But there is another way of answering the objection—by asking what should prevent Chaucer from forming opposter from *oppono*, in the same way that imposter, or impostor, has been formed from *impono?* The French have *composteur* for what we call composing-stick; and if we were to follow their example, and establish *composter*, who could demur to it? Granting, then, that Chaucer might have framed opposter, the addition of a feminine e is all my interpretation requires.

Either explanation of the word ought to be sufficient to justify its acceptance, when it furnishes such an appropriate and probable solution of this long-pending crux in Chaucer's text. The identity of expression which it would complete in all three languages is, in my opinion, a very strong argument in its favour. Finding the compound of Statius so literally reproduced by Boccaccio, it is in the highest degree probable that Chaucer would desire to shew that he could produce an exact equivalent in his own language: for that he was as intimately acquainted with Statius as he was with Boccaccio may be proved almost to demonstration by a comparison of certain identities of description common to him and Statius, but which having no existence in Boccaccio's text, could not have been derived from it.

For example, Chaucer describes the House of Mars:—

———"downward from an hille under a bente."—*Knight's Tale*, 1981.

Statius places it under Mount Hæmus, in Thrace:—

———"adverso domus immansueta sub Hæmo."—*Thebaid.* vii. 42.

Boccaccio makes no mention of hill or mountain.

Again Chaucer describes,—

"The dores were alle of Athamante eterne."—*K.T.* 1990.

Statius. ———"adamante perenni—fores"—*T.*vii. 68.

Boccaccio (in the Milan edition of the Teseide)
———"le porte a dur diamante."—vii. 32.

Chaucer. "The statue of Mars upon a carte stood."—*K.T.* 2041.

Statius. "Ipse (Mars) subit curru."—*T.* vii. 70.

Boccaccio. No corresponding expression.

Chaucer. "The cartere over ryden with his carte
Under the wheel ful lowe he lay adoun."—*K.T.* 2022.

Statius. "Et vacui currus, protritaque curribus ora."—*T.* vii. 56.

Boccaccio. "I voti carri, e li volti guastati."—*T.* vii. 37.

Here "*under the wheel*" and "*protrita curribus*" convey the same idea; but the line from the Teseide is a mere catalogue—Boccaccio speaks of damaged countenances, but he omits the cause of them.

It is needless to cite more of these parallels: enough has been shown to prove that Chaucer must have had independent recourse to the text of Statius as well as to that of Boccaccio, and, therefore, that he had a double inducement to the formation of "Shippes oppostere*s*," *as I have interpreted it.* This double reference is also a complete answer to the absurd notion that he might have read *bellatrici* as *ballatrici.*

When Chaucer represents these shippes opposteres as "*brent,*" he does not originate the burning—he only transfers it from the *incensis urbibus*" and "*terre arse*" of Statius and Boccaccio: and it cannot be denied that it was a most judicious alteration.

Contending ships mutually in flames is a fine and effective emblem of war, and one that might well be depicted in mural embellishment: but when Tyrwhitt explained "hoppesteres" as female hoppers, or dancers, and thought burning ships "dancing on the waves' would be a poetical idea, he omitted to explain how that lively motion was to be represented on the walls of the oratory: for it must be recollected that the emblematic objects which in Statius are real, in Boccaccio partly real and partly *istoriati*, are in Chaucer wholly pictorial.

In the existing text of the Knight's Tale I have always felt perplexed and dissatisfied with the frequent repetition, in the first person, of *I saw* and *saw I;* as though Chaucer had intended to make the Knight say that he had himself seen those oratories, prepared by Theseus for a temporary purpose so many ages before. This would be such a useless and unmeaning anachronism that, rather than believe it genuine, I suspect a corruption in the MS. I think we have a key to what Chaucer really did write in the preceding description of the oratory to Venus:—

> First, in the temple of Venus mayst thou see
> Wroght on the wal— &c.—1918.

May'st thou see,—that is, the Knight is presenting the picture to the mind's eye of his hearers. Now, the substitution of ye for I, in the similar descriptions of the oratories of Mars and Diana would continue the same figure of speech:—

> Ther saugh *ye* first the derk imaginyng.—1995.
> Yet saugh *ye* brent the shippes oppostere.—2017.

And so of the rest, there being no place where substitution of *ye* might not be as easily made as in these two examples: and when it is recollected that the pronoun *I* was sometimes spelled *Y* or even *Yc*, it will not appear improbable that some early scribe, perhaps Adam Scrivener himself, misread *ye* for *yc*, and under that idea copied it as *I*. It even appears that the *sound* of I was represented by the letters ye—as may be seen in "The Legende of Goode Women":—

> "The daisie, or elles the ye of day," (184).

so that the substitution of the wrong pronoun might have occurred either through the eye or the ear.

There are certain humorous touches in Chaucer's description of the House of Mars, such as:—

> The cook i-scalded for al his longe ladel—

which seem as though he were slyly quizzing the pompous descriptions of Statius and Boccaccio: and the same remark applies to his joke at their hyperbolic enumeration of the various trees cut down to construct the funeral piles, when he throws in amongst *his* trees the *whippul-tree!* or, as we might say, the axle-tree. There have been many grave discussions as to the species to which the whippul-tree belonged: a point that may perhaps be ascertained about the same time as the proper growth and culture of the axle-tree.